普通高等教育"十三五"规划教材

3D Mechanical Design and 3D Annotation Using UG NX
（UG NX 三维机械设计及三维标注）

主　编　张瑞亮
副主编　丁　华　姚爱英
　　　　辛宇鹏　刘　峰

中国铁道出版社有限公司
CHINA RAILWAY PUBLISHING HOUSE CO., LTD.

Abstract

The objective of this book is to help the readers to understand the fundamentals of computer-aided design (CAD) concepts and become familiar with the operations in CAD modeling so as to effectively use CAD approaches for mechanical product design. The commercial CAD system, UG NX is employed as the platform for CAD environment. In Chapter 1, the brief introduction to the development process of 3D modeling techniques in CAD and its applications to mechanical design are given. Following this introduction, important topics in product modeling using UG NX, including the basics of UG NX, sketch, part modeling, assembly modeling, drawing and 3D annotation are provided.

This book aims at providing a textbook for graduate students of science and technology universities whose major are Mechanical Design Manufacturing and Automation, Automobile Engineering, Mechanics, Process Equipment and Control Engineering, etc. This book provides the main guidelines for the reader to apply 3D software tools to support practical design application. Moreover, this book provides engineering senior a comprehensive reference to learn advanced technology in support of engineering design using 3D CAD technology. In addition to classroom instruction, this book should support practicing engineers who wish to learn more about the e-Design paradigm at their own pace.

图书在版编目（CIP）数据

UG NX 三维机械设计及三维标注=3D Mechanical Design and 3D Annotation Using UG NX：英文/张瑞亮主编. —北京：中国铁道出版社，2019.2（2019.12 重印）
普通高等教育"十三五"规划教材
ISBN 978-7-113-25416-2

Ⅰ.①U… Ⅱ.①张… Ⅲ.①机械设计-计算机辅助设计-应用软件-高等学校-教材-英文 Ⅳ.①TH122

中国版本图书馆 CIP 数据核字(2018)第 301029 号

书　　名：	3D Mechanical Design and 3D Annotation Using UG NX（UG NX 三维机械设计及三维标注）
作　　者：	张瑞亮　主编
策　　划：	许　璐　　　　　读者热线：（010）63550836
责任编辑：	许　璐
封面设计：	刘　颖
责任校对：	张玉华
责任印制：	郭向伟

出版发行：中国铁道出版社有限公司（100054，北京市西城区右安门西街 8 号）
网　　址：http://www.tdpress.com/51eds/
印　　刷：三河市兴博印务有限公司

版　　次：2019 年 2 月第 1 版　　2019 年 12 月第 3 次印刷
开　　本：787 mm×1 092 mm　1/16　印张：12.75　字数：254 千
书　　号：ISBN 978-7-113-25416-2
定　　价：39.80 元

版权所有　侵权必究

凡购买铁道版图书，如有印制质量问题，请与本社教材图书营销部联系调换。电话：（010）63550836
打击盗版举报电话：（010）51873659

PREFACE

With the development of science and technology and the globalization of the economy, traditional manual design is gradually being replaced by computer aided design (CAD). The application of computer technology in product design has evolved from the calculation and drawing in the past to the integration of today's 3D modeling, optimization design and simulation, which greatly shortens the design and manufacturing cycle of the product and improves the design quality.

The machinery industry plays a pivotal role in the entire industrial production process. Mechanical CAD is of great significance to promote the development of the machinery industry and the improvement of the level of science and technology. With the continuous improvement of China's manufacturing informatization level, the demand for high-level talents in the machinery industry who are familiar with both professional knowledge and 3D CAD technology is becoming more and more intense. Therefore, in the preparation of this book, we strive to reflect the contemporary mechanical 3D CAD. The basic content and development level, focusing on the basic content of mechanical CAD technology and the application of 3D CAD technology.

This book uses UG NX software as a 3D CAD platform. It mainly introduces the basic content and development history of mechanical CAD technology, UG basic knowledge, sketching, 3D modeling and typical mechanical parts modeling, assembly design, engineering drawings and product manufacturing information (PMI). This book takes the design of the planetary gear reducer as the main line. While introducing the UG command and function, it also introduces the ideas and methods of the typical parts and assembly design of the reducer. In the process of learning, readers can not only master the concept of CAD, the commands and functions of UG, but also know well about the methods and steps of engineering design using 3D CAD.

As a combination of theory and practice, the content is easy to understand. The main functional commands are explained with the operation examples step by step, so that beginners can apply learning commands to specific design. This can more effectively stimulate readers' interest in learning and improve learning effect. The book was written by the College of Mechanical and Vehicle Engineering of Taiyuan University of Technology, and Zhang Ruiliang was the editor. Zhang Ruiliang wrote Chapter 1; Ding Hua wrote Chapter 2 and Chapter 5; Yao Aiying wrote Chapter 3; Yang Tiantian wrote Chapter 4; Xin Yupeng wrote Chapter 6; Liu Feng wrote Chapter 7. The entire book was organized by Zhang Ruiliang, and Xin Yupeng reviewed the manuscript.

We are thankful to graduate students, Muhammad Yousaf Iqbal Rao and Anuj Desaihave, for their friendly help in the process of writing this book. At the same time, the book refers to a large amount of literature and materials, and we would like to express our deep gratitude to the original author. Due to the limited editorial level, there might be certain inadequacies or mistakes in the book and we sincerely hope that readers will criticize and correct.

Example files and slides needed for this book made by authors from College of Mechanical and Vehicle Engineering of Taiyuan University of Technology are available and could be obtained by author's e-mail (rl_zhang@163.com).

Written at Taiyuan University of Technology

2018-11-20

CONTENTS

Chapter 1 Introduction to CAD in Mechanical Engineering 1
   1.1 Development History of CAD Technology 1
      1.1.1 Wireframe models 2
      1.1.2 Surface models 3
      1.1.3 Solid models 4
      1.1.4 Parametric solid model 5
      1.1.5 Variational modeling technology 6
      1.1.6 Direct modeling and synchronous modeling techniques 7
   1.2 Application of CAD Technology in Mechanical Design 9
      1.2.1 Influence of 3D CAD technology on mechanical design 9
      1.2.2 Basic functions of CAD system 10
      1.2.3 CAD Data exchange standard 11
   1.3 Questions and Exercises 13

Chapter 2 Basics of NX 15
   2.1 The User Interface of NX 15
      2.1.1 Start and exit NX 12.0 15
      2.1.2 User interface 16
      2.1.3 Customizing the user interface 18
   2.2 Basic Operations of NX 20
      2.2.1 File operations 20
      2.2.2 Layer operations 23
      2.2.3 Object operations 25
      2.2.4 Mouse operation 28
      2.2.5 View operations 29
      2.2.6 Working coordinate system 37
   2.3 Exercises-basic Operations of NX 44
   2.4 Questions and Exercises 47

Chapter 3 Sketch 50
   3.1 Sketch Overview 50
      3.1.1 The Basic Concept of Sketch 50
      3.1.2 Sketch Preference 52
      3.1.3 Create Sketch Plane 53
   3.2 Sketching 54
      3.2.1 Sketch Point 54

3.2.2 Profile 55
3.2.3 Rectangle 56
3.2.4 Arc 57
3.2.5 Circle 57
3.2.6 Studio Spline 58
3.2.7 Derived Lines 59
3.2.8 Quick Trim 60
3.2.9 Quick Extend 61
3.2.10 Fillet 61
3.2.11 Chamfer 62
3.3 Constraints 63
3.3.1 Basic Concept of Constraints 63
3.3.2 Geometric Constraints 64
3.3.3 Sketch Dimensions 65
3.3.4 Show/Remove Constraints 66
3.3.5 Convert To/From Reference 66
3.3.6 Alternate Solution 67
3.4 Exercise-Sketch Example 67
3.5 Questions and Exercises 71

Chapter 4 Part Modeling 73
4.1 Feature Modeling Overview 73
4.1.1 Common modeling methods 73
4.1.2 Feature-based modeling process 75
4.1.3 Feature type 76
4.2 Boolean Operations 77
4.3 Basic Features 81
4.3.1 Extrude Features 81
4.3.2 Revolve Features 84
4.3.3 Sweep along Guide 84
4.4 Placement of Engineering Features 87
4.4.1 Placement Face 87
4.4.2 Horizontal Reference 87
4.4.3 Positioning methods 88
4.5 Datum Features 89
4.5.1 Datum Plane 89
4.5.2 Datum Axis 90
4.6 Exercises - Feature Modeling 91
4.7 Modeling of Typical Mechanical Parts 93
4.7.1 Modeling of gear shaft 93
4.7.2 Modeling of case part 99
4.8 Exercise - Cylinder Gear Shaft 108

4.9　Questions and Exercises 111

Chapter 5　Assembly Modeling 117

5.1　Assembly Overview 117
5.1.1　Basic process of NX assembly modeling 117
5.1.2　Terms in NX assembly 118
5.1.3　Introduction to the assembly interface 119
5.1.4　Assembly Navigator 120
5.1.5　Reference set 124
5.2　Assembly Method 124
5.2.1　Bottom-up assembly modeling 124
5.2.2　Top-down assembly modeling 126
5.3　Assembly Constraints 129
5.3.1　Constraint status 129
5.3.2　Fix 130
5.3.3　Touch Align 130
5.3.4　Distance 132
5.3.5　Concentric 132
5.3.6　Center 132
5.3.7　Angle 133
5.3.8　Parallel and Perpendicular 133
5.3.9　Fit and Bond 133
5.3.10　Align/Lock constraint 134
5.4　Component Edit 134
5.4.1　Mirror 134
5.4.2　Pattern 137
5.5　Exploded Views 139
5.5.1　Create exploded views 139
5.5.2　Unexplode component and Delete explosion 141
5.6　Assembly Interference Check 142
5.7　Exercise - Planetary Reducer Output Shaft Assembly 142
5.8　Questions and Exercises 147

Chapter 6　Drafting 149

6.1　Introduction to Drafting 150
6.2　Drafting Preferences 151
6.3　Drafting Management 151
6.3.1　The 1st angle projection and the 3rd angle projection 152
6.3.2　Creating a new drawing 153
6.3.3　Edit/Open/Delete sheet 155
6.4　Drafting Views 156
6.4.1　Base view 156

6.4.2 Projected view 156
6.4.3 Detail view 157
6.4.4 Section view 158
6.4.5 View break 163
6.5 Edit Drafting View 163
6.5.1 Setting dialog box 163
6.5.2 Section line and hinge line 164
6.5.3 Non-sectioned component 164
6.6 Drafting Annotations 165
6.6.1 Centerlines 165
6.6.2 Note 166
6.6.3 Geometric dimension and tolerancing in drafting 167
6.6.4 Assembly drawing 169
6.7 Exercise-Drafting Example 173
6.8 Questions and Exercises 176

Chapter 7 Product and Manufacturing Information 178
7.1 Configuration 178
7.2 Creating PMI 181
7.2.1 PMI annotation plane 181
7.2.2 PMI associated objects 181
7.2.3 Sizing PMI 182
7.3 Dimensions 183
7.3.1 PMI dimensions 183
7.3.2 Converting feature dimensions to PMI 185
7.3.3 Tips for using dimensions 185
7.3.4 PMI hole callouts 186
7.4 Annotations 187
7.4.1 Note 187
7.4.2 Feature Control Frame 187
7.4.3 Datum Feature Symbol 187
7.4.4 Datum Target 188
7.4.5 Surface Finish 188
7.4.6 Weld Symbol 188
7.4.7 Balloon 189
7.5 Section View 189
7.6 Exercise-PMI Example 190
7.6.1 Add PMI to a 3D model 190
7.6.2 Create a PMI section view 194
7.7 Questions and Exercises 195

References 196

Chapter 1 Introduction to CAD in Mechanical Engineering

This chapter aims at introducing the development history of CAD technology and its application in mechanical design, the advantages of 3D mechanical design and the data exchange standard of CAD.

The main aspects of this chapter are:

(1) Master development of CAD technology;

(2) Understand the advantages of 3D mechanical design;

(3) Understand CAD data exchange standards.

1.1 Development History of CAD Technology

The first computer graphics system was invented by the United States in the 1950s, then a passive computer-aided design techniques with simple plot output capabilities was developed. At the beginning of 1960s, surface modeling in computer-aided design (CAD) application became a reality, and commercial computer graphics equipment was introduced in the mid-1960s. In 1970s, a complete CAD system began to form. Thereafter, a raster scan display capable of producing realistic graphics appeared and various forms of graphic input devices such as manual cursors and graphic input boards were introduced that promoted the development of CAD technology. It began to grow rapidly as drawing on the computer screen became feasible. People wished to use this technology to eliminate the traditional manual sketching that is cumbersome, time-consuming and low in drawing accuracy. At that time, the starting point of CAD technology is to use the traditional three-view method to express the parts and use 2D engineering drawings produced on paper as media for technical communication. This technology is known as two-dimensional (2D) modeling technology. The meaning of CAD at this time is only a substitute for the drawing board, it means

Computer Aided Drawing (or Drafting) rather than CAD that has been discussed here. At the beginning, 2D system AutoCAD was produced by Autodesk, dominated the CAD drawing market in China. Nowadays CAD users, especially in the initial CAD users, 2D drawings still occupy a substantial proportion in China.

CAD technology has tremendous experienced development in its almost 70-years evolution history, and its technological development process is shown in Figure 1-1.

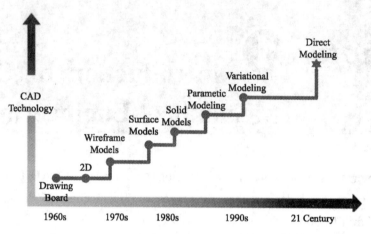

Figure 1-1   Development history of CAD technology

1.1.1 Wireframe models

In the end of 1960s, we began to study the construction of 3D solids with wireframes and polygons. The model using this form is called wireframe model. This modeling method describes objects with a set of edges formed entirely by vertices just like the frame made by the wire. Thus, the wireframe model gets its name. The wireframe model brings accessibilities for the production of engineering drawings such as a simple structure, low requirements on computer performance, can represent three-dimensional data of basic objects, can generate arbitrary views, and can maintain a correct projection relationship between views. In addition, trimetric views and perspective views can be generated, which brings a big improvement in 2D system. It is inconclusive for the reality of the object in a wireframe because all the curves are displayed. It is impossible to represent the nonpolygon surface such as a cylinder or a sphere due to the lack of surface geometric information. In addition, especially the relationship information between edges and faces, faces and faces are missing in the data structure. Because of that, we cannot construct a solid entity, cannot identify the face and body, cannot distinguish the inside or outside of a solid object. A wireframe model is not able to support a finite element mesh for structural analysis of a physical object other than beam or truss structures. The initial wireframe modeling system can only express basic geometric information and cannot effectively express the topological relationship between geometric data. Due to the lack of surface information of the model both CAM and CAE can't be realized (Figure 1-2).

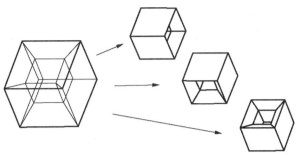

Figure 1-2　Wireframe model

1.1.2　Surface models

In the 1970s, the aircraft and automotive industries are in the growing development. There were a large number of free-form surfaces problems encountered in aircraft and automobile manufacturing. At that time, only multi-section views and characteristic parallel lines could be used to determine the free-form surfaces. Due to the incompleteness of the three-view representation method, it was often seen that the sample produced was very different or even completely different from what the designer imagined after the design was completed. Designers can't guarantee whether the shape of the surface designed can meet the requirements. Therefore, the clay model was often produced proportionally as the basis for design evaluation or scheme comparison. The slow and complicated production process had greatly delayed the development time of the products, and the requirement for updating the design means was getting higher and higher. In the 1970s, the Bézier algorithm was proposed by Pierre Bézier, which provided the designer to deal with surface and curve problems with a computer (Figure 1-3). The surface modeler CATIA was released by Dassault Systemes in 1981. A surface can be thought of as an infinitely thin shell stretched over a wireframe. Based on the data structure of the wireframe model, surface models represent a shape by its surface geometry. It is formed by adding various related data that can form the solid surface. A surface model includes information about the faces and edges of a part and can express the topology between the edge and the face. In surface model we can achieve face-to-face intersection, rendering, surface area calculation, hiding and other functions. The application of the surface model indicates that CAD technology is freed from the three-view mode that simply acts like the engineering drawings. First time the main information of the product parts is fully described by the computer. A surface model is good for visualizing complex surfaces and supports NC tool path generation. The surface modeling system CATIA brought the first CAD technology revolution. The development methods of complex products such as airplanes and automobiles are better than the old ones. The design development cycles has also been greatly accelerated and the automotive industry has begun to use CAD technology in large quantities.

Since the surface model can only represent the surface and boundary of the object and it cannot be cut. The physical properties such as mass, centroid, and moment of inertia of a solid object represented in a surface model is hard to determine, and it is difficult to express complex manufacturing information. Additional information must be added to a surface model in order to specify in/out and top/bottom of the physical object that the surface model represents.

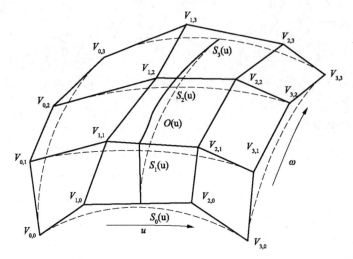

Figure 1-3 Bézier surface

1.1.3 Solid models

The problem of CAM can be basically solved with the surface model. However, since the surface model technology can only express the surface information of the shape, it is difficult to accurately express other characteristics of the part such as mass, center of gravity and moment of inertia etc., which are very unfavorable to CAE. The biggest problem is that the pre-processing of the analysis is particularly difficult. Based on the exploration of the development of CAD/CAE integration technology, Integrated Design and Engineering Analysis Software (I-DEAS) based on solid modeling technology was produced by Structural Dynamics Research Corporation in 1982, and used primarily in the automotive industry, most notably by Ford Motor Company and General Motors. Solid models contain information about the edges, faces, and the interior of the part. The mathematical description contains information that determines whether any location is inside, outside, or on the boundary surface. In terms of mathematically representing a solid object, two major modeling methods, constructive solid geometry (CSG) and boundary representation (B-rep), are widely employed by geometric modeling kernels, which are the core of CAD systems. CSG approach expresses the sequential process of modeling, and B-rep model specifies connectivity information of vertex, edge, face, and volume that must be defined to construct the solid model. The solid modeling technology can accurately express all the attributes of the part, and contain information about 3D entity of the part. It gives guarantee that only the parts that can actually be realized are shaped will not lack edges, faces, and no edge will penetrate into the part entities, thus avoiding errors and unrealizable designs. An advanced overall shape definition method can be provided to support new models from old models through Boolean operations. The solid model theoretically helps to bring together model expressions in CAD, CAE and CAM which brings amazing convenience to the product design.

A solid model is the ultimate way to represent general objects, which are physically solid objects. It represents the future way of CAD technology. Based on this consensus, it became the mainstream of CAD technology development at that time. It could be said that the popular application of solid modeling technology marked the second technological revolution in the history of CAD development (Figure 1-4).

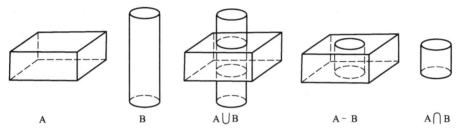

Figure 1-4 Schematic diagram of the union, difference, and intersection of a 3D primitives

Solid modeling technology not only brings the improvement of the algorithm and the hope of future development but also brings the extreme expansion of the amount of data calculation. Under the hardware conditions at that time, the calculation and display speed of solid modeling was very slow and it was rather reluctant to design in practical applications. CAE based on the physical model is originally a higher-level technology that's why the popularity is cramp and the reflection is not strong. In addition, in the face of the contradiction between algorithm and system efficiency, many companies give favor solid modeling technology that do not have the power to go. Develop it, but instead turn to the surface model technology that is relatively easy to implement. The technical orientation of each company has once again divided. The solid modeling technology has not been fully promoted in the entire industry.

1.1.4 Parametric solid model

Parametric is a term used to describe a dimension's ability to change the shape of model geometry as soon as the dimension value is modified. In the mid of 1980s, for the problems of unconstrained free-form modeling techniques, the researchers proposed a better algorithm than the unconstrained free-form modeling, called parametric solid modeling method. In 1989, Parametric Technology Corporation (PTC) developed Pro/ENGINEER software using object-oriented unified database and fully parametric modeling technology that providing an excellent platform for 3D solid modeling. Parametric modeling manipulates parameters to control the geometric shape of a solid objects. Parameters come from dimensions in 2D profiles in sketch, dimensions on 3D solid features and variables in user-defined equations. The core of parametric modeling is to use geometric constraints, engineering equations and relationships to illustrate the shape characteristics of the product model. If defined properly, the entire part geometry can be controlled by a small number of key parameters. Design intents can therefore be captured through the change of the small set of parameters. The dimensions of the geometric entities that make up the product shape can be parametrically related as equations. The parametric modeling technique is also called the dimension-driven geometry technology. It brought the third technological revolution in the history of CAD development.

The guiding thought of the parametric system is only need to operate in the way specified by the system. The system gives guarantee of correctness and efficiency of the design that you generate otherwise it refuses to operate. This kind of operation also has a lot of side effects. First of all, the user must follow the intrinsic use mechanism of the software such as never allowing underconstraint and solving in reverse order etc. Secondly, it is difficult for the designer to put all the dimensions when the cross-sectional shape of the part is complicated. It is difficult to determine which dimensional modification will satisfy the design intents. Finally, the dimension-driven range is also limited, if an unreasonable parameter is given to cause a feature to interfere with other features, a change in topological relationship is caused. Therefore, from the application point of view, the parameterization system is particularly suitable for the parts industry where the technology is quite stable and mature. In such an industry, the shape of the part changes little, and often only the analog design is used that is the shape is basically fixed, and only a few key dimensions need be changed to obtain a new serial design result.

1.1.5 Variational modeling technology

Parametric technology requires that all the dimensions are fully defined, that is the designer must consider the shape and size together in the early design stage and the whole design process. It controls the shape through the size constraint and drives the shape modification through the dimension change. Dimensional parameters are regarded as a starting point. Once the shape of the designed part is too complicated, how to change these sizes to achieve the desired shape is not intuitive. Moreover, while topological relationship of key shapes are changed in the design, geometric features losing certain constraints can also cause system data to be confused. In fact, full constraint is a rigid rule for designers which will interfere with the creativity and imagination of the designer.

In fact, when we are doing mechanical design and process design, we always hope that the parts can be built as we desire and can be dismantled at will, allowing us to construct three-dimensional designs on the computer screen. We hope to keep each intermediate result for use in iterative design and optimal design. In response to this demand, SDRC developers have proposed a more advanced variable technology, variational geometry extended (VGX), based on parametric technology.

The variable technology further distinguishes the size "parameters" defined in the parametric technique into geometry constraints and size-constraints, rather than constraining all geometries by size only as in the parametric technique. The reason for adopting this technology is that in the conceptual design phase of a large number of new product developments, the designer first consider the design ideas and then fixes them in certain geometric shapes. The exact dimensions and the strict dimensional positioning relationships between shapes are difficult to be fully determined at the initial stage of design, so it is naturally desirable to allow the existence of under size-constraints in the initial stages of the design. In addition to considering the geometry constraints the variable design can also directly solve the engineering relationship as a constraint directly with the geometric equation without additional model processing. The advantages of using variable technology are:

(1) The designer can adopt the design method of the size after shape, allowing the use of incomplete size constraints and only giving the necessary design conditions. In this case, the correctness and efficiency of the design can still be guaranteed.

(2) The modeling process is a process similar to an engineer thinking about a design in his mind. The geometry that meets the design requirements is the first and the dimensional details are gradually and subsequently improved.

(3) The design process is relatively free and relaxed. The designer can consider the design scheme more and does not need to care too much about the inherent mechanism of the software and the design rules. Therefore, the application field of the variable system is also vast.

(4) In addition to the general serialized part design, the variable system is especially handy when doing conceptual design, and is more suitable for innovative design such as new product development and old product remodeling design.

Based on the idea of variational modeling theory, SDRC launched I-DEAS Master Series software in 1993, formed a set of unique variational modeling theory and software development methods. The variational technology not only maintains the original advantages of parametric technology but also overcomes many of its disadvantages. Its successful application has provided more space and opportunities for the development of CAD technology and also driven the fourth technological revolution of CAD development. At present, most CAD systems provide support for variable technology, and variabilization is also the mainstream of current 3D CAD systems.

1.1.6 Direct modeling and synchronous modeling techniques

The core of solid modeling is constructive solid geometry (CSG) and boundary representation (B-rep) modeling method. CSG expresses the sequential process of modeling and B-rep is the vertex, edge, face, and body information of the 3D model. The feature-based parametric modeling system adds the concept of feature tree to CSG. This is the kernel principle of popular mainstream feature-based 3D mechanical CAD systems at this time.

The direct modeling core has only B-rep information but doesn't have CSG information, because the order of the modeling is not considered, so designers can quickly define and edit geometry by simply clicking on the model geometry and moving it. In 2008, Siemens PLM Software took the lead in releasing synchronization technology in the PLM industry. The 3D modeling technology was further improved forming a variety of modeling methods including direct modeling, feature modeling, surface modeling and synchronization technology. The synchronization technology is a three-dimensional modeling method and constraint solving technique that combines feature-based modeling and direct modeling to achieve dimension-driven and stretch deformation (Stretch) in a three-dimensional environment. Direct modeling method can retain the solid feature information of the part and implement the design changes. Thus, the history-less modeling and feature-based parameter modeling are perfectly compatible, enabling rapid modification of the 3D model, helping achieve rapid design changes and product family design.

Synchronous technology advances a large step on top of existing history-based parametric modeling technique. Synchronous modeling technique examines the current geometric relation of the product model in real time and combines them with the parameters and geometric constraints added by the designer. Thus, the designer can evaluate and build new geometric models, and edit the model without having to repeat the entire modeling history. Designers no longer have to research and analyze complex constraint relationships to understand how to edit models. They don't have to worry about subsequent model associations for editing. Synchronous modeling technology breaks through the inherent architectural barriers of history-based design systems. The drawbacks of the current design process caused by parametric modeling technology can be avoided. Designers can effectively perform dimension-driven direct modeling and no re-creation or conversion is required. Synchronous modeling technology can modify the product more quickly, which will greatly improve design efficiency and reduce product development costs, therefore shorten the time to market (Figure 1-5).

Figure 1-5   Synchronous modeling

According to the design task and the product characteristics of the design, the designer should determine what kind of modeling method will be employed. For example, the art modeling uses the surface model to flexibly represent various complex surface features. Parametric modeling system that supports specific parametric, feature and history-based modeling emphasizes the integrity and systematicness of the design ensuring that any design changes will update all models. This can achieve more advantages in terms of capturing and reusing design intents, family-based or platform-driven product design. It is more convenient to optimize the design as the parametric feature-based modeling retains the design history. The synchronous modeling approach has higher flexibility and so in the aspects of rapid design, prototype design, etc. has more advantages.

1.2 Application of CAD Technology in Mechanical Design

1.2.1 Influence of 3D CAD technology on mechanical design

Compared with traditional mechanical design, CAD technology has great advantages in terms of increasing productivity, improving design quality, reducing costs and reducing labor intensity, mainly as follows:

(1) CAD can improve design quality. The comprehensive technical knowledge is stored in the computer system, which provides a scientific basis for product design. The interaction between computers and users is conducive to combine the power of human being and computers, and making product design more deliberate. The optimized design method used by CAD contributes to the optimization of certain process parameters and product structures. In addition, different departments can use the information in the same database and data consistency is maintained.

(2) CAD can save time and increase productivity. Automation of design calculations and modeling greatly reduces design development time. The integration of CAD and CAM can significantly shorten the cycle time from design to manufacturing, and the design efficiency can be increased by 3 to 5 times compared with the traditional design method.

(3) CAD can reduce costs by a large margin. The high-speed operation of the computer and the automatic operation of the plotter save labor force greatly. At the same time, optimized design brings savings in raw materials. Some of the economic benefits of CAD can be estimated, while others are difficult to estimate. Due to the adoption of CAD/CAM technology, the production preparation time is shortened, and the product replacement is accelerated which greatly enhances the competitiveness of the product in the market.

(4) CAD technology allows designers to focus on more creative work rather than doing massive calculations and drawing work. In the product design process, the manually created drawings workload accounts for about 60% of the total workload. In CAD, this part of the work is done by the computer and the benefits are very significant.

(5) CAD/CAE/CAM integration: CAE is an important module of 3D CAD software.

CAE functions include engineering numerical analysis, structure optimization design, strength design evaluation and life prediction, dynamics, kinematics simulation and so on. After completing the product modeling in CAD system, the CAE module can analyze the rationality, strength, stiffness, fatigue, material, structural rationality, motion characteristics, interference, collision and dynamic characteristics of the design. CAE technology has also been widely used in China. Taking the automobile manufacturing industry as an example, many domestic automakers and automobile design companies use CAE analysis after completing the design of new models using 3D CAD software. Such as interference inspection, sheet metal forming analysis, plastic part draft angle analysis, body strength stiffness test and the motion simulation function in the CAE module are widely used in moving parts such as windows, doors, wipers, etc., to calculate the movement trajectory of the parts, and the state of the parts in motion. In this way, CAE can provide an intuitive reference for designers. These analyses have greatly improved the reliability of new models, shortened the product development cycle, reduced rework and saved R&D costs. With 3D CAD technology, mechanical design time has been reduced by nearly 1/3. At the same time, the 3D CAD system has a highly variable design capability, so a brand new mechanical product can be obtained through rapid reconstruction, which greatly improves the work efficiency.

Because CAD technology has the advantages of simplicity, efficiency, high precision and convenient storage, it has been widely used in many fields such as machinery, aerospace, shipbuilding, construction, electronics, etc. It has achieved fruitful results and huge economic benefits. CAD technology fundamentally changes the quality of mechanical design, overcomes many of the drawbacks of traditional mechanical design and makes mechanical products have a qualitative change.

(1) CAD technology is more innovative than traditional mechanical design. Traditional mechanical design focuses on practical experience and design methods are mostly based on analogy, retrofit design and lack of innovation. When using CAD technology for mechanical design, it can have a strong three-dimensional effect which can stimulate the designer's innovative thinking and design a new more creative product.

(2) CAD technology lays the foundation for the mechanical design informatization. CAD technology fully computerizes the entire mechanical design process. Thus, the design defects and deficiencies can be found in time through the three-dimensional design technology before the specific processing and manufacturing, and can be improved in time, which not only shortens the design time but also improves the quality of the design.

1.2.2　Basic functions of CAD system

CAD system is composed of system software, supporting software and application software. Although different CAD systems can have different functional requirements, in terms of mechanical 3D CAD systems, at least the following basic functions should be included:

(1) Geometric modeling: The geometric modeling of the product is the core of the CAD system. The CAD, CAE, CAM and other subsequent processing tasks are all carried out on the basis of the geometric model. The strength of the geometric modeling largely reflects the strength of CAD system, most of the mainstream CAD systems now provide the support for geometric modeling techniques such as wireframe modeling, surface modeling, feature-based parametric solid modeling, and direct modeling.

(2) Assembly modeling: Assembly modeling is the core module of 3D mechanical CAD. The basic design idea of 3D CAD is based on the mechanical manufacturing process. That is to design only the parts and then to assemble the parts into components by assembly function, and further to assemble them into a product. It is relatively simple to complete the simple part design of these structures since the part is the smallest unit that can be manufactured in the machine. By assembling these basic parts into products, it can be achieved that not only greatly simplifies the design complexity but also conforms to the mechanical manufacturing process.

(3) Engineering drawing module: Nowadays 3D CAD has become the mainstream but 2D engineering drawings also are widely used. Therefore, generating 2D engineering drawings according to 3D models becomes a basic function of 3D CAD system.

(4) Scientific calculation and analysis functions: it can carry out the routine and optimum design of the product. It can perform scientific calculations on reliability, finite element analysis, dynamic analysis and digital simulation.

(5) Data management and exchange functions: such as database management, data exchange and interfaces between different CAD systems.

(6) Other functions: CAD has such as document production, editing, word processing functions, software design functions and network functions, etc.

1.2.3   CAD Data exchange standard

With the continuous development, maturity of CAD technology and the continuous improving of CAD applications in various industries, CAD standardization work has increasingly shown its importance. As part of the high-tech standardization, CAD standardization occupies a very important position in the CAD technology. The State Standardization Administration and the State General Administration of Quality Supervision, Inspection and Quarantine jointly issued the "*Specification for CAD general technology*", which specifies all general aspects of CAD technology in China. CAD data exchange problem is an important issue faced by various industries after the widespread application of CAD. Due to the rapid expansion of CAD data, the data files generated by different CAD systems use different data formats, and even the types of data elements in different CAD systems are not the same. This situation potentially hinders the further application and development of CAD technology. Therefore, how to accomplish the sharing and effective management of CAD information in enterprises is a very important issue for standardization.

At present, the most commonly employed neutral files for CAD model translations mainly are DXF (Data Exchange File), IGES (Initial Graphics Exchange Specification) and STEP (Standard for The Exchange of Product model data) application protocols.

DXF file is an ASCII text file with a special format. It is easy to read and easy to be processed by other programs. AutoCAD is widely used worldwide, so this data file format has become a de facto industry standard. The DXF data exchange file provides great convenience for the promotion of CAD/CAM technology. However, the DXF file is developed earlier, from the current point of view, it certainly has many shortcomings. It cannot describe the complete geometric model of the product. It is difficult to further develop because it is less in support of solid model translation.

The IGES (Initial Graphics Exchange Specification) project was started in 1979 by a group of CAD users and vendors, including Boeing, General Electric, Xerox, Computer Vision, and Applicon, with the support of the National Bureau of Standards (now known as NIST) and the U.S. Department of Defense (DoD).Soon after, it was adapted and recognized by American National Standard Institute as a standard tool format. Consequently, IGES has become an acceptable and widely used neutral format for translator development by many CAD/CAM software vendors. After the initial release of STEP (ISO 10303) in 1994, interest in further development of IGES declined, and Version 5.3 (1996) was the last published standard. IGES has been used in the automotive, aerospace, and shipbuilding industries. These part models may have to be used years after the vendor of the original design system has gone out of business. IGES files provide a way to access this data decades later. However, IGES unreasonably defines a direct access pointer system in the file structure. The main problems exposed in the application are: (1) the data file is too large, (2) the data conversion processing time is too long, (3) some geometric type conversion is unstable, (4) only pay attention to the graphics data conversion and ignore the conversion of other information. Nevertheless, IGES is still the de facto international standard data exchange format currently widely used in many countries. China has adopted IGES 3.0 as the national recommendation standard since September 1993.

The work with the ISO 10303 standard, informally called Standard for the Exchange of Product model data (STEP) was initiated in 1984 with the goal to standardize exchange of product data between product life cycle systems. The standard is a very comprehensive set of specifications, covering the entire life cycle of the product such as design, analysis, manufacturing, testing, inspection, etc. Using STEP-supporting tools, data such as geometry, topology, tolerances, relationships, attributes and performance can be exchanged. In addition, some data related to processing may be included. The product model provides comprehensive information for release production tasks, direct quality control and testing. STEP provides a unique description and computer processing form of information representation throughout the life of the product. This form is independent of any particular computer system and guarantees consistency across multiple applications and systems. This standard also allows for different implementation techniques to facilitate access, transfer and archiving of product data. The STEP standard is a common resource and application model developed for the provision of neutral product data for CAD/CAM systems. It covers all product areas of architecture, engineering, construction, machinery, electrical engineering and building structure, etc. In terms of product data sharing, the STEP standard provides four levels of implementation as follows: (1) ASCII neutral files, (2) application interfaces for accessing memory structure data, (3) shared databases, (4) shared knowledge library. Undoubtedly, this will bring a big change in the business and manufacturing industry.

The STEP standard has obvious advantages in the following aspects: (1) the economic benefits are significant, (2) the data range is wide, the precision is high and the application protocol eliminates the ambiguity of product data, (3) easy to integrate and to expand, (4) advanced technology and clear hierarchy, divided into six parts such as general resources (40-series), application resources (100-series) and application protocols (200-series), etc.

Today, the STEP standard has become an internationally recognized global standard for the exchange of CAD data files. Many countries have developed corresponding national standards based on the STEP standard. The corresponding standard number of STEP standard in China is GB16656. The problem with the STEP standard is that the entire system is extremely large. The standard development process is slow, and the data files are larger than IGES format files. At present, the STEP application protocol provided by commercial CAD system only has AP203 (configuration-controlled 3D designs of mechanical parts and assemblies) and AP214 (core data for automotive mechanical design processes).

1.3 Questions and Exercises

1. Fill in the Blanks

(1) 3D modeling technology mainly includes wireframe models, _____models, _____models, _____models, _____technology and _____technology.

(2) For mechanical 3D CAD systems, at least the following basic functions should be available basic functions, _____ functions, _____ functions, _____ functions and _____ functions.

2. Short answer questions

(1) Please briefly describe the advantages of 3D CAD?

(2) What are the main applications of CAD technology in the machinery industry?

(3) What are the main CAD data exchange standards?

Chapter 2

Basics of NX

NX (referred to as UG) is a Product Lifecycle Management (PLM) software produced by Siemens. It integrates functions such as CAD/CAE/CAM and covers the whole process from concept design to product production. It provides solid modeling, surface modeling, assembly modeling, drawings and other functions commonly used in 3D CAD software. NX also provides product validation functions such as model interaction verification, simulation and manufacturing.

As a beginner of NX, being familiar with NX interface and basic operation methods is the basis for learning the software and the key to improve the application ability of the software.

This chapter will introduce the user interface, view operations, mouse operations, layer operations and coordinate system operations, etc. to help you get started on learning and using NX.

The main points of this chapter are:
(1) Be familiar with the user interface of NX;
(2) Understand the related methods of view layout;
(3) Usage and skills of the model orientation view and rendering style;
(4) Transformation method of the working coordinate system;
(5) Methods of file operation and layer setting.

2.1 The User Interface of NX

2.1.1 Start and exit NX 12.0

When the NX 12.0 is started, its startup interface as shown in Figure 2-1. The startup interface will disappear after a while, then the NX 12.0 initial operation interface will be shown as Figure 2-2.

Figure 2-1　NX 12.0 startup interface

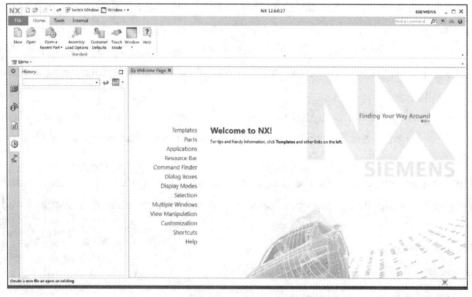

Figure 2-2　NX 12.0 initial operation interface

In the NX 12.0 initial operation interface, it provides brief information about *Templates*, *Parts*, *Applications*, *Resource bar*, *Command Finder*, *Dialog Boxes*, *Display Modes*, *Selection*, *Multiple Windows*, *View Manipulation*, *Customization*, *Shortcuts* and *Help*. It will be very helpful for beginners to carefully review these information in getting started with NX.

To exit NX 12.0, click the *Close* button on the right side on the *Title* bar of NX 12.0.

2.1.2　User interface

Click the *New* button to create a new file in the NX 12.0 initial operation interface, or click the *Open* button to open the model file, you can enter the main operation interface of NX 12.0. Figure 2-3 shows the main operation interface for the modeling.

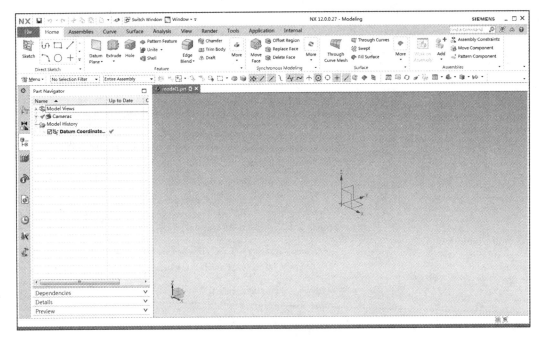

Figure 2-3　NX 12.0 main operation interface

The main operation interface mainly includes *Title* bar, *Quick Access* toolbar, *Ribbon* bar, *Resource* bar, *Graphics* area, *Cue/Status* bar and *Top Border* bar (including menu buttons, selection bars, view toolbar, and application toolbar). Among them, the *Quick Access* toolbar is initially embedded in the *Title* bar, which is used to display and collect some common tools for users to quickly access the corresponding commands.

(1) *Title* bar: displays information such as the NX version, the module currently selected by the user, the file name of the current working part, and the modification status of the current working part.

(2) *Ribbon* bar: displays the commonly used functions of NX. It provides a shortcut icon button for menu commands. A command is represented by a corresponding icon, and a series of icons with similar functions form a toolbar.

(3) *Cue/Status* bar: located below the graphics area. It is mainly used to prompt you for the next action and display message. The cue line lists guiding information related to the function you are performing. When the mouse is over a graphic element, the status bar displays the name of the current graphic element; when the user has selected some certain graphic elements, the status bar will display the currently selected element name or quantity.

(4) *Resource* bar: including user interfaces such as *Navigator*, *Reuse Library*, *History*, and *Roles* tab. The corresponding user interface can be opened by clicking the icon in the resource bar.

(5) *Graphics* area: displays the model with which you are working on. It occupies most of the space of the screen and is used to display the result of the user operation.

Table 2-1 describes the functions of the common button on the Resource bar.

Table 2-1　Common button function description on the Resource bar

Name	Icon	Description
Assembly Navigator		Through the assembly structure tree of the graphic display component, the assembly process and related constraints of the component can be seen, and the edit operation method of the component in the assembly is provided
Part Navigator		Display the feature structure tree of the part, from which you can see the part creation process and related parameters, and provide the edit operation method of the features in the part
Reuse Library		Provide rapid design of standard parts, common or reusable parts, no need to re-create parts when assembling, just load the relevant part name from the reuse library. By automatically selecting or modifying the part parameters, the needed parts will be loaded into the assembly
History		Provide direct access to recently opened files, which can be opened by clicking and dragging the file to the graphics workspace
Roles		Display the system default and user-defined role types. Usually, the user interface settings of the corresponding roles are saved under the role. The user can load the corresponding user interface by clicking the corresponding role type

2.1.3　Customizing the user interface

When working with NX 12.0, sometimes you need an adequate large graphics window, sometimes you need to add or remove some toolbars on the operation interface, etc. This involves the NX interface customization issue. Here are a few practical tips related to customizing the user interface.

1. Enable function area tab

When working with some application modules, sometimes the function area only provides the common tabs associated with the task instead of enabling the existing tabs. If necessary, the user can enable one of the other tabs for more efficiency. For example, to enable the *Application* tab in the function area, right-click the empty area of the function area and select the tab option you want to enable in the popup shortcut menu, for example, select the *Application* option.(Figure 2-4)

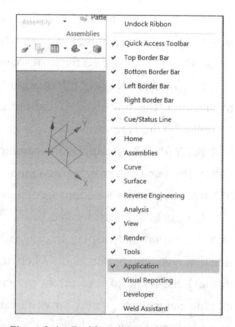

Figure 2-4　Enable a ribbon tab by right-click

2. Use the *Customize* command

In the Tools menu, select the *Customize* command, or right-click anywhere on toolbar and choose the *Customize* command from the shortcut menu. The system will show the *Customize* dialog box, which can be used to customize the menu and toolbar, icon size, screen display, cue/status line location, and more. For example, in the *Tabs/Bars* tab as shown in Figure 2-5, check or clear the check box in front of the toolbar tab name to enable or disable the selected the toolbar. In the *Icons/Tooltips* tab, you can set the icon sizes of some specific toolbars and set display option of tooltips, as shown in Figure 2-6.

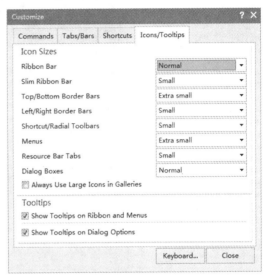

Figure 2-5　*Tabs/Bars* tab of the *Customize* dialog box　　Figure 2-6　*Icons/Tooltips* tab of the *Customize* dialog box

In NX 12.0, if you want to add a tool command to a group in a specified toolbar or a tab of a ribbon, switch to the *Commands* tab in the *Customize* dialog box. Select a category from the *Categories* list to display all the commands under this category. Then select the desired command in the *Items* list box, as shown in Figure 2-7, and drag this command to the specified toolbar or a tab group in the function area to finish its placement. In addition, the operation of customizing menu options are similar.

3. Load roles

Roles tab displays the system default and user-defined role types. Usually, the user interface settings of the role are saved under the

Figure 2-7　*Commands* tab of the *Customize* dialog box

corresponding role. The user can load the corresponding user interface by clicking the corresponding role type.

While learning NX, you may want to use a limited set of its tools. You can achieve this by choosing an appropriate role from this category that tailors the user interface, thereby hiding the tools and commands you do not need for day to day tasks. Click the *Roles* tab on the *Resource* bar to browse the system default roles, and click the desired role to load corresponding set of tools as shown in Figure 2-8.

Figure 2-8　Load role in the *Resource* bar

2.2　Basic Operations of NX

2.2.1　File operations

The basic operations of file management commonly used in NX 12.0 mainly include creating new files, opening files, saving files, closing files, importing and exporting files, and so on.

1. Creating a new file

Press the <Ctrl+N> keyboard shortcut, or choose *File* tab>*New*, or click the *New* button in the *Quick Access* toolbar. The *New* dialog box will appear as shown in Figure 2-9. The dialog box provides 14 tabs for creating 14 types of files. Here, the *Model* tab is used as an example to show how to create a model file.

(1) At this time, the status bar displays the timing information of "Select a template and, if required, a part to reference". Switch to the *Model* tab and select the *Units* option from the *Units* drop-down list box, such as *Millimeters*, *inches or all*.

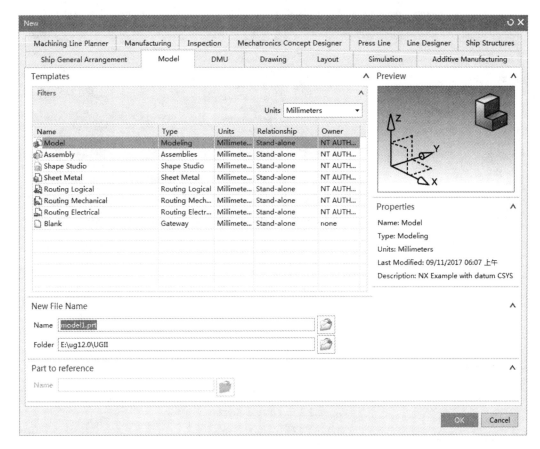

Figure 2-9   *New* file dialog box

(2) Type a new name or accept the default name in the *Name* text box of the *New File Name* option group. The file name can usually consist of English letters and Arabic numerals, but it cannot contain Chinese characters. Since NX uses the same extension (*.prt) for types such as *model* and *drawing*, it usually adds the corresponding type name to the file name to distinguish between different types of files, such as: for *Modeling*, will add "_model" to the file name; for *Assemblies*, will add "_asm" to the file name; for *Drawing*, will add "_dwg" to the file name, etc.

(3) Select folder: specifies the directory where the new file is located. You can enter a new directory, or browse for a directory. The folder name cannot contain Chinese characters.

2. Open a file

To open an existing file, press the <Ctrl+O>, or choose *File* tab>*Open*, or click the *Open* button in the *Quick Access* toolbar to bring up the *Open* dialog box as shown in Figure 2-10. Browse the directory and select the file to open. If necessary, set whether to load the settings. You can preview the file to be opened by checking the *Preview* check box and click the *OK* button.

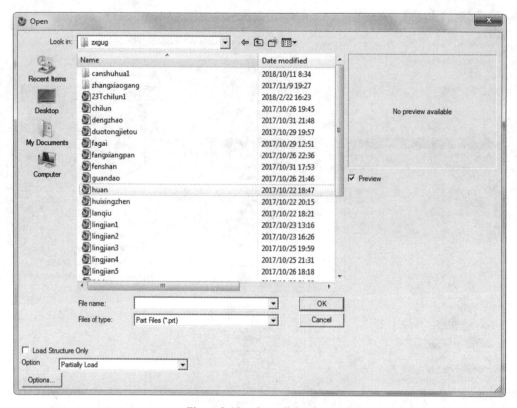

Figure 2-10　*Open* dialog box

3. Save and save as a file

NX provides 5 methods for saving files as shown in the Table 2-2.

Table 2-2　Five kinds of save operation commands

No	Command	Description
1	Save	Save work parts and any modified components
2	Save Work Part Only	Save only the working parts
3	Save As	Save the current working part in the specified directory with a different name
4	Save All	Save all modified parts and all top assembly parts
5	Save Bookmark	Save assembly associations in bookmark files, including component visibility, load options, and component groups

After modifying the newly created or opened file, you can select *File>Save* or click the icon on the toolbar to save the file to the original directory; if you need to save the current file as another defined name or save to other directories, you can select the menu command *File>Save As*. *Save As* dialog box is shown in Figure 2-11. The user can specify the file save directory in *Save in*, enter the new file name in *File name* and specify the file save format by selecting the *Save as type* option and then click the *OK* button to save the file.

4. Close file

To close the file, choose *File>Close* and select the appropriate option to perform the close operation. The user can also close the current window by clicking the button in the upper right corner of the graphics window. When the window is closed, the system will prompt whether to save the modified file.

Figure 2-11 *Save As* dialog box

2.2.2 Layer operations

Layers are designed to make it easier for users to manage design models. Layers are similar to transparent drawings. Drawing objects on a working layer is like drawing an object on a specified piece of transparent paper. Users can change the working layer according to the design status, and can set layers which are visible and which are invisible. Objects can be displayed or hidden through layers without affecting the spatial position and interrelationship of the model, and the displaying, editing, and state of objects can be controlled during complex modeling. Layers are used to store objects in a file, and work like containers to collect the objects in a structured and consistent manner. In NX, layers are represented and distinguished by layer numbers, and layer numbers cannot be changed. Up to 256 layers can be set in each model file, which are represented by layer numbers 1 to 256.

1. Layer settings

There are four types of layer states, namely the work layer, the selectable layer, the visible only layer, and invisible layer, as listed in Table 2-3.

To set the layer, choose *Menu Format>Layer Settings*, the *Layer Settings* dialog box will appear as shown in Figure 2-12. Use this dialog box to set the working layer, visible layer and invisible layers, and can define the category name of the layer, and so on.

In order to facilitate the layer selection, the *Show* option can be used to control the layer category and name displayed in the layer list. The *Show* list includes 4 options: *All Layers* means that all layers are displayed in the layer list; *Layers With Objects* means that only the layers containing objects are displayed in the layer list; *All Selectable Layers* means that only the selectable layers are displayed in the layer list; *All Visible Layers* means that only visible layers are displayed in the layer list.

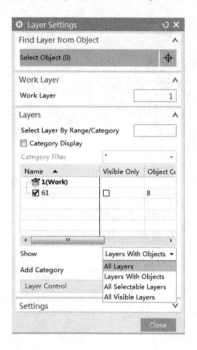

Table 2-3 Layer state

State	Description
Selectable	Make the layer optional. Objects on this layer can be displayed in the graphics workspace and can be selected for any subsequent operations. Objects on selectable layers are also visible
Work layer	Set the layer as a working layer. This way each object created later will be on that layer. Objects on the working layer can be displayed in the graphics workspace and can be selected for any subsequent operations
Visible only	Make the layer visible. Objects on this layer can be displayed in the graphics workspace, but cannot be selected
Invisible	Make the layer invisible. Objects on this layer can not be displayed in the graphics workspace, and objects on the invisible layer cannot be selected

Figure 2-12 *Layer Settings* dialog box

2. Move to layer

If you don't make layer settings before creating the model or put some irrelevant geometric elements in an unsuitable layer, in this case, you can use the *Move to Layer* function to move the element from currently used layers to destination layer.

To move an object to destination layer, the user can show the *Class Selection* dialog box by selecting the menu command *Format>Move to Layer* or click the icon on the toolbar. After selecting an object, click *OK* button to open the *Layer Move* dialog box. To identify the destination layer, you can enter the layer name or number in the *Destination Layer or Category* text box or select the corresponding layer in the *Layer* list box, click *OK* button to move the selected object to the specified layer as shown in Figure 2-13.

To move more other objects, you can click the *Select New Objects* button in the *Layer Move* dialog box. At this time, you will return to the *Class Selection* dialog box and repeat the above operation steps to complete moving the new object.

Chapter 2 | Basics of NX | 25

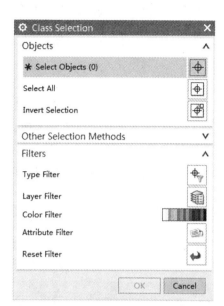
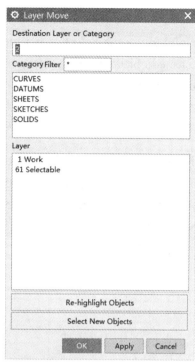

Figure 2-13   *Move to Layer* operation process

2.2.3 Object operations

In NX, in order to improve work efficiency and object operation accuracy, the user needs to perform operations such as object selection, object display setting, and object showing and hiding.

1. Object selection

During the design work, selecting object is an important basic operation. You can move the mouse pointer over the object and click the left mouse button to select the object. Repeat the operation to select other objects. To cancel selected object, hold down the <Shift> key and click the object or press <Esc> key.

When multiple objects are close together, you can use the *QuickPick* dialog box to select the desired object. The method is to place the mouse pointer on the object to be selected and wait until the 3 points appear next to the mouse pointer, click to open the *QuickPick* dialog box as shown in Figure 2-14, which lists multiple objects under the current mouse pointer. Move cursor to browse object which will be highlighted, and then click to select the object. When using the *QuickPick* dialog box, you can use the relevant buttons in the dialog box, such as *All Objects*, *Construction Objects*, *Features*, *Body Objects*, etc. These buttons play the role of selecting filter, for example, when you click the *Body Objects*, only the related body objects are listed in the object list for the user

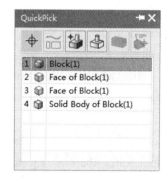

Figure 2-14   *QuickPick* dialog box

to select.

It is helpful to use the *Selection* bar toolbar (Figure2-15) located in the top border bar to match the selected object, because the selection filter can be specified in the *Selection* bar to set the selection method.

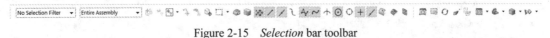

Figure 2-15   *Selection* bar toolbar

Selection Filter: it can be used to specify the type of the selected object, as shown in Figure 2-16. After the user specifies the particular object type, only objects belonging to that type can be selected in the graphics workspace.

Selection Scope: it is used to filter the selection to a portion of the displayed model as shown in Figure 2-17.

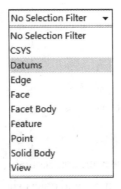
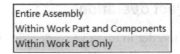

Figure 2-16   Select filter   Figure 2-17   Selection range

Snap Point: By enabling the *Snap Point* option, the specific control points on objects such as curves, edges, and faces can be selected in the graphics workspace.

In addition, the user can select the object to be selected in the Navigator structure tree.

2. Edit object display

Edit object display refers to modifying the display attribution of selected object, such as layer, color, line font, width, transparency and analysis display options. The specific steps are as follows:

Select the menu command *Edit>Object Display* or press the shortcut key <Ctrl+J> to bring up the *Class Selection* dialog box. Select the desired object in the graphics area and click *OK* button to bring up the *Edit Object Display* dialog box as shown in Figure 2-18. In the dialog box, the user can edit the options in the *General* and *Analysis* tabs. After setting the options, click the *OK* button to complete the edit object display. The specific meanings and setting methods of main options in the *General* tab are shown in Table 2-4. The effect of before and after color and transparency setting is shown in Figure 2-19.

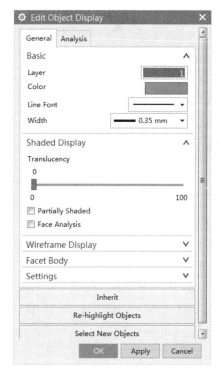
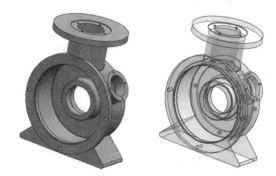

Figure 2-18 *Edit Object Display* dialog box   Figure 2-19   *Effect of Edit Object Display* option setting

Table 2-4   Meaning and setting methods of the main options in the General tab

Option name	Meaning and setting method
Layer	Be used to specify the layer in which the object is located. Set the layer where the current object is located by entering the layer number
Color	Be used to set the color of the object. The user clicks the color bar behind the *Color* tab, and the *Color* dialog box is displayed. In the dialog box, you can select the color that the object needs to be displayed
Line Font	The line font is used to set the edges of the object. Object line font setting by selecting the corresponding option in the drop-down box after the *Line Font* tab
Width	Used to set the line width of the edge of the object. Set the object line width by selecting the corresponding option in the drop-down box after the *Width* tab
Transparency	Be used to set the transparency of the object. The user drags the slider and the transparency changes as the slider moves. When the transparency is 0, the object is opaque; when the transparency is 100, the object is completely transparent

3. Show and hide object

When creating a complex model, because the model includes many objects, it is easy to cause problems such as difficulty in selecting objects, not easy to observe object and the model display speed is slow. At this time, the user can use this function to temporarily hide the object that is not currently operated, and after the corresponding operation is completed, the hidden object is redisplayed as needed.

Select the menu command *Edit > Show and Hide > Show and Hide* or press the shortcut key <Ctrl+W> to show the *Show and Hide* dialog box as shown in Figure 2-20. This dialog box controls the show or hide state of all objects in the current graphics workspace. The various object types contained in the current model are collected in the *Type* column of the dialog. Show or hide this type of object can be done by clicking the button in the *Show* column to the right of the type name or the button in the *Hide* column.

Figure 2-20 *Show and Hide* dfialog box

2.2.4 Mouse operation

The function of the mouse in NX is very powerful. The function of the mouse will be different when the user is in different operating states. The commonly used mouse has three buttons, they are left mouse button (MB1), middle mouse button or scroll wheel (MB2), and right mouse button (MB3). The basic functions of the commonly used mouse buttons are as follows:

1. Menu management

(1) Left mouse button: in the graphic workspace or dialog box, click MB1 to select object; double-click MB1 on an object to activate the default operation of this object, such as to show the properties of the feature object, the component object displayed as working part or the like.

(2) Middle mouse button: be used to finish selection, in the dialog box, click the MB2 is equivalent to click the *OK* button; hold down the <Alt> key and click MB2, which is equivalent to click the *Cancel* button in the dialog box; MB2 also can be used to complete all the necessary steps step by step during the modeling process.

(3) Right mouse button: right click on the object to display the shortcut menu of the object; in the graphic workspace, click MB3 to display the View operation shortcut menu and selection bar.

2. View operation

(1) Middle mouse button: in the graphics area, scroll the mouse wheel for zooming the view; hold down the middle mouse button and drag the mouse to rotate the view, as shown in Figure 2-21(a). When using the middle mouse button to rotate the view, the system provides different rotation modes depending on the position of the cursor in the graphics work area, as follows:

When the cursor is in the graphics work area, drag the middle mouse button to rotate the object in the view around the geometric center of the model or the rotation point, as shown in Figure 2-21(b).

To set or clear the rotation point, click the right mouse button in the blank space of the graphic work area, select *Set Rotation Reference* in the shortcut menu as shown in Figure

2-21 (c), and specify a point as the rotation point in the graphics workspace. When the user sets the rotation point, the same operation can be performed to *Clear Rotation Point*.

When the cursor is near to the border of the graphic work area, press the middle mouse button, the cursor shape listed in Table 2-5 will appear, and drag the middle mouse button to rotate model around the corresponding axis, as shown in Figure 2-21(d).

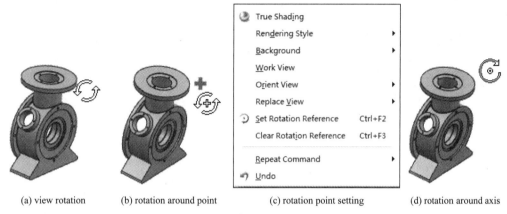

(a) view rotation   (b) rotation around point   (c) rotation point setting   (d) rotation around axis

Figure 2-21 Mouse rotation operation

(2) <Shift> + middle mouse button: press the <Shift> key and the middle mouse button simultaneously in the graphics work area, and drag the mouse to pan the model.

Table 2-5 Single-axis rotation cursor shape

Cursor shape	Cursor position	Cursor description
	Graphic work area left or right edge	Rotate around the X axis
	Lower border of the graphics workspace	Rotate around the Y axis
	Upper border of the graphics workspace	Rotate around the Z axis

(3) <Ctrl> + middle mouse button: in the graphics work area, press the <Ctrl> key and the middle mouse button at the same time, drag the mouse to zoom in/out the model to gain a better understanding of the model geometry and its constituent components.

(4) Right mouse button: in the blank area of the graphic work area, press and hold the right mouse button to show the shortcut menu, as shown in Figure 2-22, you can view and render the view.

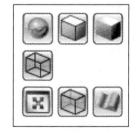

Figure 2-22 Radial shortcut menu

(5) Press the middle and right mouse buttons simultaneously to pan the view.

2.2.5 View operations

When designing with NX, users often need to observe the model from different directions, that is, observe the model view by changing the angle of view. In addition, in different design environments, in order to express the model structure more clearly, it is necessary to change the display mode of the model, that is, display the model in a wireframe or shaded manner. That is, the work of modeling is inseparable from the operation of the view.

The view operation commands used are located in the *Operation* cascading menu under *View* in the *Menu* bar, as shown in Figure 2-23. The following is a brief introduction to the meaning of several major view operation commands, so that the reader can get a preliminary understanding.

Figure 2-23　View operation command

① Refresh: during the user's work, some temporary prompts may be displayed on the screen or the screen may not be updated immediately. At this time, the user can click the *Refresh* button to force the screen display to be updated to eliminate the temporary prompt.

② Fit: adjust the center and scale of the work view to display all objects.

③ Zoom: zoom in and out of the working view.

④ Origin: change the center of the working view.

⑤ Pan: when this command is selected, the view is panned by pressing the left mouse button and dragging the mouse.

⑥ Rotate: use the mouse to rotate the view around a specific axis.

⑦ Orient: orient the work view to the specified coordinate system.

⑧ Restore: restores the working view to the orientation and scale before the last view operation.

1. Orient view

1) System predefined view

NX provides eight default views of the model viewed from a particular direction, as shown in Figure 2-24. The eight orientation views are trimetric, isometric, top, bottom, front, back, left, and right. The user can change the viewing direction of the model by selecting the corresponding view from the *Orientation View* option or the *shortcut* menu in the *View* toolbar. The user can also change the current view of the model to the trimetric view by using the *Home* button. *End* button change the current view of the model to the isometric view, and the <F8> key changes the current view of the model to the other 6 plane views that are closest to the current view. It should be noted that the direction reference of these predefined views takes absolute coordinate system as reference, which is not a directional view in the true engineering sense.

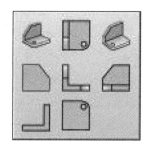

Figure 2-24 Orient view type

(1) Trimetric: observe the model from the right-front-up direction of the coordinate system, as shown in Figure 2-25(a).

(2) Isometric: observe the model from the right-front-up direction of the coordinate system in an equiangular relationship, as shown in Figure 2-25(b).

(a) Positive second measurement view   (b) Positive isometric view

Figure 2-25 Measurement view

(3) Top: a view projected onto the XC-YC plane in the negative ZC direction, as shown in Figure 2-26(a).

(4) Bottom: a view projected onto the XC-YC plane in the positive direction of ZC, as shown in Figure 2-26(b).

(5) Left: a view projected in the positive direction of XC onto the YC-ZC plane, as

shown in Figure 2-26(c).

(6) Right: a view projected onto the YC-ZC plane in the negative XC direction, as shown in Figure 2-26(d).

(7) Front: a view projected onto the XC-ZC plane in the positive direction of YC, as shown in Figure 2-26(e).

(8) Back: a view projected onto the XC-ZC plane in the negative direction of YC, as shown in Figure 2-26(f).

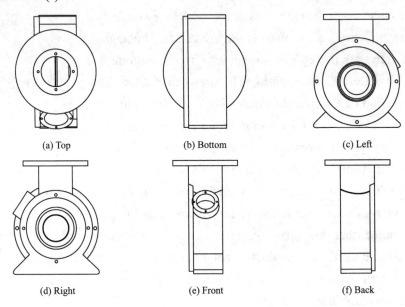

(a) Top   (b) Bottom   (c) Left

(d) Right   (e) Front   (f) Back

Figure 2-26 Basic view

2) Custom view

Since the predefined view provided by the NX is not the direction of the model view desired by the user in some cases, the user can observe the model and the subsequent operations through the customized view.

Select the menu command *View>Operation>Orient*, and the *CSYS* dialog box will pop up. In this dialog box, complete the setting of the coordinate system according to the user's needs, click the *OK* button, and the model will display according to the defined view direction. Then select the menu command *View>Operation>Save As*, and the *Save Work View* dialog box will pop up, as shown in Figure 2-27. Input a new view name in the dialog box and click the *OK* button to save the new view of the model. The new view can be called by double-clicking the name of the custom view in the *Model Views* in the *Part Navigator*; you can also select the right-click menu command *Orient View>Custom Views* of the graphic workspace, double-click the name of the custom view to call it in the *Orient View* dialog box as shown in Figure 2-28.

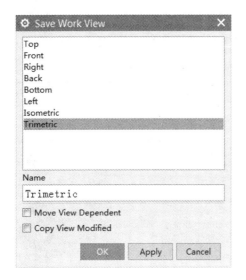
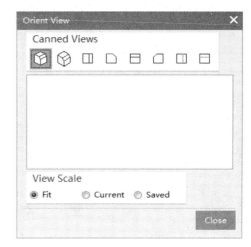

Figure 2-27 *Save Work View* dialog box   Figure 2-28 *Orient View* dialog box

2. Rendering style

The model rendering styles commonly used in NX are divided into two categories: shaded and wireframe display. The shaded display is divided into two types: shaded and shaded with edges. The wireframe display is divided into static wireframe, wireframe with hidden edges, and wireframe with dim edges. The user can set the display style in the rendering style drop-down list on the top *View* toolbar. As shown in Figure 2-29.

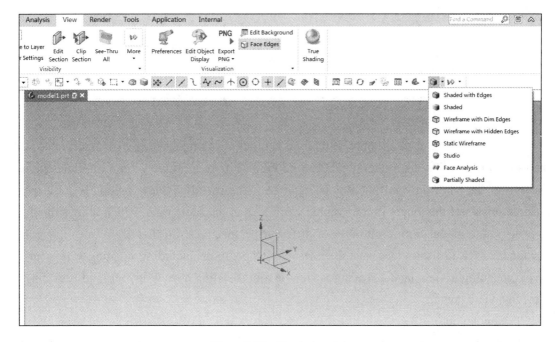

Figure 2-29 Rendering style

The functions and effects of common rendering styles are shown in Table 2-6.

Usually, in the model sketching environment, the *Wireframe with Hidden Edges* is used to display the style, and the *Shaded* display style is applied in the modeling environment.

Table 2-6　Functions and effects of common rendering styles

Style name	Icon	Features	Effect
Shaded with Edges		The model shades and displays the edges of the model	
Shaded		Model coloring display, does not display the edges of the model	
Static Wireframe		Model wireframe display, hidden edge normal display	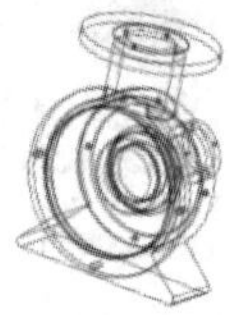
Wireframe with Hidden Edges		Model wireframe display, hidden edges are not displayed	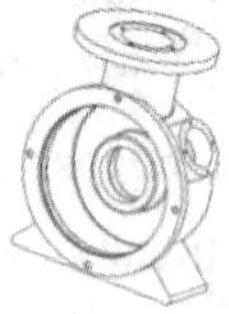
Wireframe with Dim Edges		Model wireframe display, hidden edge fade display	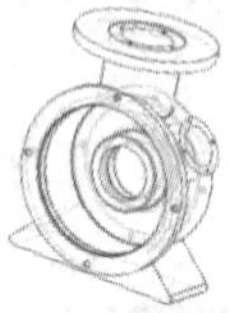

3. View layout

When designing a 3D product, it may be applied to multiple views of a model object at the same time, so that the model object can be viewed more intuitively from multiple angles. Therefore, the view layout function is required. The so-called view layout is to display different views of the model in the graphic work area at the same time, which allows the user to observe the model from a plurality of different perspectives, So that users can fully grasp the model. NX 12.0 provides useful view layout operation, including creating a new view layout, opening a view layout, saving a view layout, deleting a view layout, and replacing a view in a view layout.

1) New layout

Before applying the view layout, users need to create a new view layout according to their needs. Select *Menu>View>Layout>New* menu to open the *New Layout* dialog box, as shown in Figure 2-30.

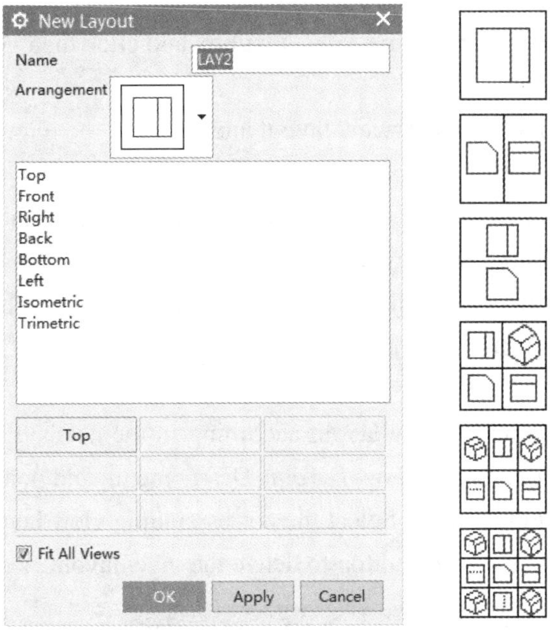

Figure 2-30   *New Layout* dialog box

Input the name of the new layout in the *Name* text box of the *New Layout* dialog box, then select the corresponding layout type in the *Arrangement* drop-down list. The NX system provides six layout formats, and up to nine views can be placed to view the model.

After determining the *Arrangement* format, the *layout view* button at the bottom of the dialog box will be activated, and the number of active view buttons will be automatically updated with the selection of the layout type. If 4 view layouts are selected, 4 view buttons are displayed. The user can accept the default view layout of the system, or specify the position of each view in the layout according to his own needs: firstly select the view to be transformed by clicking the *view* button, then select the corresponding view in the *view* list box.

After the user completes the selection of the layout format and view type, click the *OK* button to complete the layout creation. The new view layout is shown in Figure 2-31.

2) Save layout

In order to call the newly created view layout in a later view operation, save it after a new layout is created. The layout is saved in two ways: by the original name of the layout and by saving as another layout name.

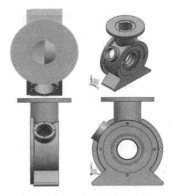

Figure 2-31   Model view layout

The user can save the layout by its original name through the *Menu>View>Layout >Save* menu. You can also open the *Save Layout As* dialog box via the *Menu>View>Layout>Save As* menu, as shown in Figure 2-32. All the layout names that already exist, which are listed in the dialog box. Input the layout name in the *Name* text box and click the *OK* button.

3) Open layout

After the user has created a new view layout and saved it, the relevant view layout can be called by opening the layout.

Select *Menu>View>Layout>Open* menu, and the *Open Layout* dialog box will pop up, as shown in Figure 2-33. The saved layout names are listed in the list box of the dialog box (except for the currently open view layout). Select the desired layout name and click the *OK* button to complete the layout.

4) Delete layout

Users can delete redundant view layout according to the needs of design. The operation method is as follows: select *Menu>View>Layout>Delete* menu, and pop up the *Delete Layouts* dialog box, as shown in Figure 2-34. Select the corresponding view layout from the layout list box of the dialog box. Click the *OK* button to delete the view layout.

Figure 2-32　*Save Layout As* dialog box

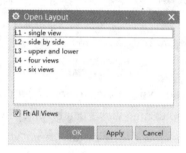
Figure 2-33　*Open Layout* dialog box

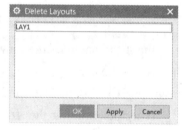
Figure 2-34　*Delete Layouts* dialog box

5) Layout view operation

(1) Work view

In a multi-view layout, the user can rotate, pan, and other views on each view, but the user can only edit the model in one view, which is the working view. In a multiview layout, there is only one working view.

You can set the layout view to work view by following methods:

In the graphics workspace, right-click on the desired view and select *Work View* from the pop-up menu.

Select *Menu>View>Operation>Select Work*, pop up the *Select Work View* dialog box, double-click the view name in the list box or select the view name and click the *OK* button.

(2) Replacement view

The user can replace any view in the saved layout if needed during the work process. The operation method is as follows: select *Menu>View>Layout>Replace View*... menu, pop up the

Replace View with dialog box, select the view to be replaced and the replacement view in the view list, and click *OK* button, as shown in Figure 2-35.

The user can also right-click the view to be replaced in the graphics workspace, select *Replace View...* in the pop-up menu, then select the replacement view.

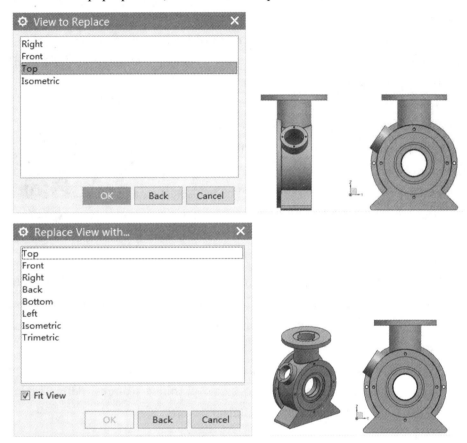

Figure 2-35   Replacement view

2.2.6　Working coordinate system

In 3D CAD software, the coordinate system is the basic reference for determining the spatial pose of each element in the 3D model, and is also the basis for view transformation and geometric transformation. In the process of creating a 3D model of the product with using NX, the user can create a coordinate system or edit and rotate the existing coordinate system if needed to make the determination of the model data parameters relative to the coordinate system more convenient, so improve design efficiency.

　　1. Type of coordinate system

In NX, the coordinate system includes three coordinate systems: absolute coordinate system, working coordinate system (WCS), and datum coordinate system (datum CSYS). The direction of these coordinate systems conforms to the right-hand rule. The absolute coordinate system is the default coordinate system of the system. It exists when the model file is created. The origin position and the direction of each coordinate axis remain unchanged, which ensures that the positional relationship of each part of the model is determined. The absolute coordinate system is used to define a fixed point and direction in the three-dimensional space, which can be used as the reference for component assembly. The working coordinate system is a movable coordinate system, and the user can move its position as needed; the reference coordinate system can provide references when creating additional features and locating components in an assembly.

1) Absolute coordinate system

There is only one absolute coordinate system in the model. In NX, the view's triad is used to represent the orientation of the absolute coordinate system. The view triad is displayed in the lower left corner of the graphics workspace. You can select an axis of the triad and then drag the middle mouse to rotate the model around the axis. As shown in Figure 2-36.

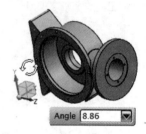

Figure 2-36   Triple axis

2) Working coordinate system

There can be multiple coordinate systems in each model, however only one of them is the working coordinate system. The working coordinate system is an important reference in the modeling process. It cannot be deleted. Users can create, edit, display and hide the coordinate system if needed.

The XC-YC plane of the working coordinate system is called the working plane. Users can refer to the work coordinate system to create primitive features, define sketch planes, create datum axes or datum planes, or create linear pattern.

3) datum coordinate system

The reference coordinate system is a reference feature, so it exists as a feature of the model and is displayed as a feature in the component navigator. It consists of three axes, three planes, a coordinate system, and an origin. These objects can be selected individually to provide reference for creating other features or positioning components in the assembly.

2. Creation of the working coordinate system

The working coordinate system is usually created by creating a datum coordinate system feature, that is, the datum coordinate system is created firstly, then the working coordinate system is moved to the datum coordinate system. The user can continue to create the original model on the newly created work coordinate system.

To create a reference coordinate system, you can open the *Datum CSYS* dialog box by selecting the *Menu>Insert>Datum/Point>Datum CSYS* menu. As shown in Figure 2-37.

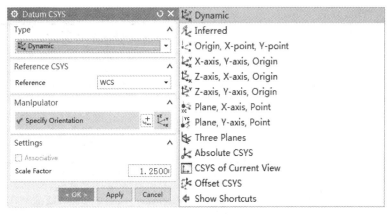

Figure 2-37 *Datum CSYS* dialog box

In the dialog box, select the method of constructing the reference coordinate system in the *Type* list box, then set other options according to the contents of the dialog box to complete the creation of the datum coordinate system. The methods for constructing various datum coordinate systems are shown in Table 2-7. These construction methods are the same as the principle of geometrically constructing the coordinate system.

Table 2-7 Methods for constructing various reference coordinate system

Method name	Method icon	Constructor description
Dynamic		Activate the current working coordinate system and move or rotate it arbitrarily; or select the existing reference coordinate system CSYS as a reference and offset it to create a new reference coordinate system
Inferred		The system automatically selects possible coordinate system construction methods based on the reference object selected by the user. When the requirements of the coordinate system construction are met, the system will automatically create a new coordinate system
Origin,X-point,Y-point		Determine 3 points in the graphics workspace to construct a new coordinate system. The order of selection of the three points is the origin, the point on the X-axis, and the point on the Y-axis. The direction in which the first point points to the second point is the positive direction of the X axis, and the direction in which the first point points to the third point is the positive direction of the Y axis
X-axis,Y-axis, Origin		Select a point and two vectors in the graphics workspace to construct a new coordinate system. One of the selected points is the origin of the coordinate system, and the other two vectors are used to determine the positive direction of the X and Y axes, respectively
Z-axis, X-axis, Origin		Select a point and two vectors in the graphics workspace to construct a new coordinate system. One of the selected points is the origin of the coordinate system, and the other two vectors are used to determine the positive direction of the Z-axis and the X-axis, respectively
Z-axis, Y-axis, Origin		Select a point and two vectors in the graphics workspace to construct a new coordinate system. One of the selected points is the origin of the coordinate system, and the other two vectors are used to determine the positive direction of the Z-axis and the Y-axis, respectively

Continued

Method name	Method icon	Constructor description
Plane, X-axis, Point		Select a plane in the graphics workspace, a vector and a point to construct a new coordinate system. One of the selected planes is the XOY plane of the coordinate system, that is, the normal plane of the Z-axis. One vector is used to determine the forward direction of the X-axis, and one point is used to specify the origin of the coordinate system
Three Planes		Select three planes in the graphics workspace to construct a new coordinate system. The normal direction of the first plane selected is the X-axis, and the normal of the second surface is the Y-axis. The third surface is used to determine the position of the origin, and the Z-axis is generated according to the right hand rule
Absolute CSYS		Construct a new coordinate system at the absolute coordinate position (0,0,0)
CSYS of Current View		Construct a new coordinate system using the orientation of the view in the graphical workspace. The XOY plane is parallel to the screen, the X-axis is horizontal to the right, the Y-axis is vertical, and the origin is in the center of the graphic work area
Offset CSYS		Referring to the existing coordinate system, a new coordinate system is constructed by inputting the offset distance and the rotation angle with respect to the X, Y, and Z coordinate axes of the existing coordinate system

After completing the creation of the reference coordinate system, select the reference coordinate system to be used as the working coordinate system, then click the right mouse button, select *Set WCS as the Reference CSYS* in the pop-up menu, and the working coordinate system will move to the reference coordinate system position. At the end, complete the creation of the working coordinate system.

3. Transformation of the working coordinate system

In the modeling process, in order to easily create the various components of the model, users can translate, rotate, redirect, change directions for new or existing coordinate systems. To change the coordinate system, the user can select *Menu>Format>WCS* menu, then select the corresponding transformation method in the pop-up submenu to perform the corresponding transformation operation. The specific methods are as follows:

1) Dynamic

Dynamics is the most commonly used transformation tool in the working coordinate system. It can translate or rotate the working coordinate system directly in the graphic work area. The user can activate the working coordinate system through the *Dynamics* menu (as shown in Figure 2-38) or double-click WCS in the graphics work area. The activated WCS is shown in Figure 2-39. The user can cancel the activation of the WCS by clicking the middle mouse button.

There are three types of operating handles on the activated WCS, which are the spherical origin handle, the spherical rotating handle and the arrow translate handle. The user can perform WCS translation, rotation, etc. by selecting and operating these handles.

(1) Origin translation

After selecting the origin handle, the user can translate the origin of the WCS to an endpoint on the model geometry or move the origin to the specified position.

(2) Translation along the coordinate axis

The user selects the translation handle corresponding to the XC, YC, and ZC axes, then drags the mouse to the mouse or enters the corresponding distance value in the input text box to move the coordinate system in the corresponding direction. When dragging with the mouse, the default distance increment is 10.

(3) Rotating around the coordinate axis

The user selects the corresponding spherical rotation handle in the XC-YC, YC-ZC, and ZC-XC coordinate planes, then drags the mouse or inputs the corresponding angle value in the input text box to rotate the coordinate system around the corresponding coordinate axis. When dragging with the mouse, the default angular increment is 45°.

(4) Reorient coordinate system

The user selects the XC, YC, ZC axes or their corresponding moving handles, and then selects the corresponding object (such as an edge) with the mouse, so that the selected axis can be parallel to the referenced object, or double-click the selected axis to flip it. Reorienting the coordinate system does not move the coordinate origin.

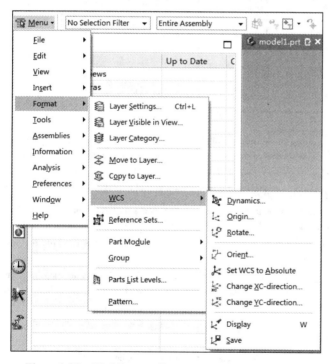

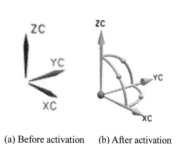

(a) Before activation (b) After activation

Figure 2-38 Working coordinate system change menu

Figure 2-39 Working coordinate system

2) Origin

Origin is used to move the origin of the current working coordinate system. The direction of each coordinate axis of the moved coordinate system does not change. After the user selects the *Origin* menu (as shown in Figure 2-38), the *Point* dialog box is displayed, as shown in Figure 2-40. The user can specify a new origin in the *Point Location* in the dialog box, or using the *Output Coordinates* select the coordinates of the input point to locate the new coordinate origin.

The methods for constructing various points are shown in Table 2-8.

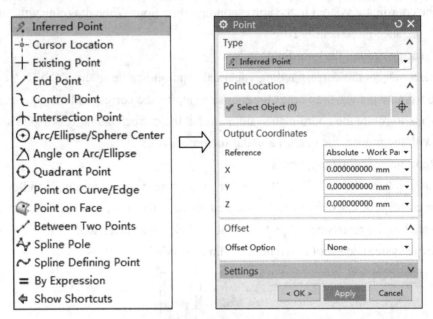

Figure 2-40 *Point* dialog box

Table 2-8 The methods for constructing various points

Method name	Method icon	Constructor description
Inferred Point		According to the location of the cursor, the system automatically captures the existing key points on the object (such as endpoints, intersections, etc.), constructs a point at the captured point, which contains the selection of all points
Cursor Location		Construct a point by locating the current position of the cursor
Existing Point		Create a new point at an existing point, or specify the location of the new point by an existing point
End Point		Construct a point with the position of the end point on the edge curve of a line, arc, spline, etc.
Control Point		Construct a point with the midpoint and end point of the line, the endpoint of the quadratic curve, the midpoint, end point, and center of the arc, or the end point, pole, etc. of the spline
Intersection Point		Construct a point with reference to the intersection of a curve and a curve, or a line-to-surface
Arc/Ellipse/Sphere Center		Construct a point at the center of the selection arc, ellipse, or ball
Angle on Arc/Ellipse		Construct a point on an arc or elliptical arc at an angle to the positive direction of the coordinate axis XC
Quadrant Point		Construct a point at the quarter of a circle or ellipse
Point on Curve / Edge		Set a point on the characteristic curve or edge line to construct a point

Continued

Method name	Method icon	Constructor description
Point on Face		Construct a point by setting the U-direction parameter and the V-direction parameter on the feature surface
Between Two Points		Determine the position of the new point by specifying a point between the two points with a two-point distance percentage
By Expression	=	Specify a point using a point type expression

3) Rotation

Rotation is to rotate the current WCS by selecting the direction of an axis of the current working coordinate system and specifying the corresponding angle. After the user selects the *Rotate* menu (as shown in Figure 2-41), *the Rotate WCS about...* dialog box is displayed, as shown in Figure 2-41. The user specifies the rotation mode in the dialog box and enters the angle value to complete the coordinates. The rotation of the system.

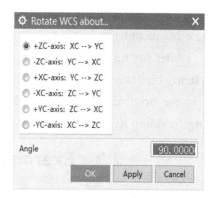

Figure 2-41 *Rotate WCS about...* Dialog

4) Orient

Orient is to position the current working coordinate system to the new coordinate system through the options in the *CSYS* dialog box. The specific operation method is the same as the *work coordinate system creation* introduced in the previous section, and will not be described here.

5) Set WCS to Absolute

Set WCS to Absolute moves the current working coordinate system to the orientation and position of the absolute coordinate system, and the coordinate direction is the same as the absolute coordinate system.

6) Change direction

The direction of change includes two options: *Change XC-Direction* and *Change YC-Direction*, which respectively reposition the orientation of the current working coordinate system by changing the direction of the X or Y axis of the coordinate system.

Select the corresponding option and pop up the *Point* dialog box. After the user determines the position of the point, the system will use the connection between the origin of the working coordinate system and the projection of the point in the XC-YC plane as the X axis or the Y-axis direction of the new working coordinate system, the direction of the Z-axis of the working coordinate system remains unchanged.

4. Display of coordinate system

Display is used to show or hide the current working coordinate system. This command is a menu command with a check box. When the check box is selected, the current working coordinate system is displayed; when the check box is cancelled, the current working coordinate system is hidden.

5. Coordinate system save

Save is used to save the current working coordinate system. Usually, the coordinate system created after transformation needs to be saved in time, so that it can be distinguished from the original coordinate system, and it is also convenient for the user to call at any time during the subsequent work.

After saving the current working coordinate system, the saved coordinate system changes from the original XC, YC, and ZC axes to the corresponding X, Y, and Z axes, as shown in Figure 2-42.

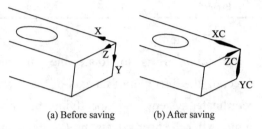

(a) Before saving   (b) After saving

Figure 2-42　Saving the coordinate system

2.3　Exercises-basic Operations of NX

This section introduces a basic entry example of NX 12.0, which aims to deepen understanding of the basic operations of NX.

The specific steps of this example are as follows:

(1) After starting NX 12.0, click the *Open* button in the *Quick Access* toolbar to open the supporting file "NX_CH.prt" in this chapter, as shown in Figure 2-43.

Figure 2-43　Opening the model file

(2) Place the mouse pointer in the drawing window, hold down the middle mouse button and move the mouse to flip the model view to the view effect shown in Figure 2-44.

(3) Press <Ctrl+Shift+Z>, or select *Menu>View>Operation>Zoom* command. The system will pop up the *Zoom View* dialog box shown in Figure 2-45 and click *Reduce 10%* button, then click the *Half Scale* button, pay attention to the effect of view zoom, and then click the *OK* button.

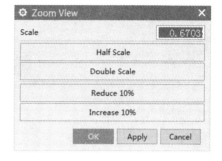

Figure 2-44 Flip model view display   Figure 2-45 *Zoom View* dialog box

(4) In the graphics window, drag the mouse while holding down the middle and right mouse buttons to practice panning the model view.

(5) Right-click in the blank area of the graphics window and select the *Orient View>Isometric* command from the pop-up shortcut menu, or click the *Isometric* button in the *Layout* group of the *Functional Area View* tab, and locate the working view to align with the isometric map. The model display effect is shown in Figure 2-46.

(6) Right-click in the blank area of the graphics window and select the *Orientation View>Trimetric* command from the pop-up shortcut menu so that the working view is positioned to align with the positive triaxial drawing, and the model display effect is shown in Figure 2-47.

Figure 2-46 Isometric view   Figure 2-47 Positive triaxial view

(7) Press <Ctrl+Shift+N>, or click the *Menu>Layout>New* button in the View tab of the ribbon, and the *New Layout* dialog box will pop up. Input *LAY6 in the Name* text box, and select the layout mode option as "L4" as shown in Figure 2-48. Then click the *OK* button in

the *New Layout* dialog box. The result is shown in Figure 2-49.

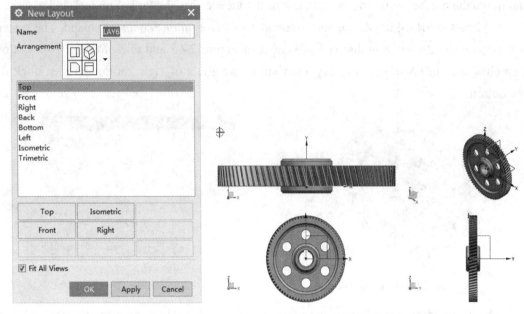

Figure 2-48 *New layout* diagram

Figure 2-49 New layout result

(8) Click the *Undo* button in the *Quick Access* toolbar to undo the last operation. In this example, the last new layout operation is undone.

(9) Press the right mouse button on the blank area of the view of the graphics window for 1 second to open the view radial menu. Hold the right mouse button and move the mouse crosshair to the *Wireframe with Dim Edges* button. As shown in Figure 2-50, when the right mouse button is released, the button option can be selected. The spur gear part will be displayed in a wire frame with dim edges. The effect is shown in Figure 2-51.

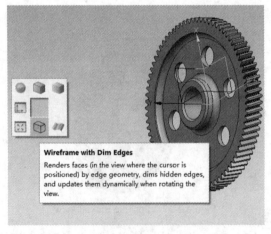

Figure 2-50 Opening the view radial menu and selecting options from it

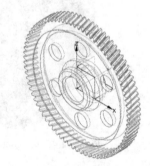

Figure 2-51 Wireframe display with dim edges

2.4 Questions and Exercises

1. Fill in the blanks

(1) In the NX working interface, _____ used to display the NX version, the working module currently selected by the user, the file name of the current working part, and the modification status of the current working part.

(2) The navigator is divided into _____ and _____, which is a tree for displaying and managing the operation of the current component.

(3) In the _____ layers of NX, the control of the layer is divided into four states, namely, _____, working state, _____ and _____. Multiple layers can be used when designing a part, and there are _____ work layers.

(4) When creating a complex model, because the model includes a large number of objects, it is easy to cause problems such as difficulty in selecting objects when the user operates, objects are not easy to observe, and the model display speed is slow. At this time, the user can _____ to temporarily hide the object that is not currently operated, and after the corresponding operation is completed, the hidden object is redisplayed as needed.

(5) When setting the default save path of a new file in NX, the path cannot contain _____.

2. Choice

(1) In NX, _____ the default coordinate system of the system. It exists when the model file is created. The position of the origin and the direction of each coordinate axis remain unchanged. This ensures that the positional relationship of each part of the model is determined. It can be used as a benchmark for component assembly.

  A. Absolute coordinate system   B. Working coordinate system
  C. Reference coordinate system   D. Characteristic coordinate system

(2) Use the mouse to observe the object. When the cursor is in the graphic workspace, drag the middle button to rotate the object in the view around the geometric center or rotation point of the model. In the graphic workspace, press the _____ button and the middle mouse button simultaneously. Move the mouse to pan the view.

  A. <Ctrl>   B. <Shift>   C. <Tab>   D. <Alt>

(3) NX will automatically add a new file name according to the template type selected by the user. The file name cannot contain _____ .

  A. Capital letter   B. Lowercase letter
  C. Arabic number   D. Chinese character

(4) When performing view layout, the NX system provides _____ layout format, and up to _____ view can be arranged to observe the model.

  A. 5, 8   B. 6, 8   C. 6, 9   D. 7, 9

(5) When applying a model rendering style, the display style is usually applied in _____ sketching environment.

  A. Coloring   B. Wireframe with hidden edges

  C. Static wireframe   D. Edged coloring

3. Short answer questions

(1) Briefly describe the composition of the working interface of the NX software.

(2) List the methods of working coordinate system transformation.

(3) Briefly describe the view operation function of the mouse.

(4) How to change the layer where the object is located?

4. Exercises

(1) Open ch2\exercise\2-1.prt and customize the toolbar as shown in Figure 2-52. The specific requirements are as follows:

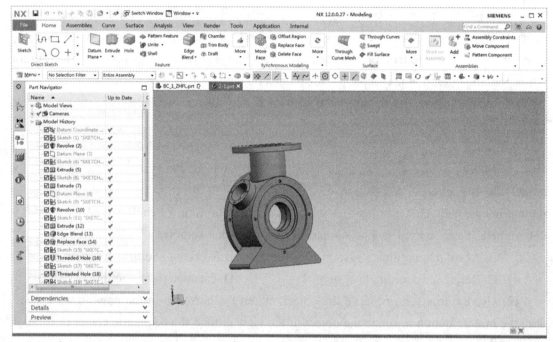

Figure 2-52 Custom toolbar

① 3 work bars including standards, selection bars and features;

② The shortcut button displayed by the feature toolbar is as shown.

(2) Open ch2\exercise\2-2.prt and perform layer, view and coordinate system operations. The specific requirements are as follows:

① As shown in Figure 2-53 (a), moving the stretch feature from layer 1 to layer 20 and using layers to display and hide it;

② Select the upper surface as shown in Figure 2-53 (b), transform the view to the direction parallel to the plane, and then apply the various rendering styles to view the model;

③ Change the coordinate system from the position shown in Figure 2-53 (b) to the position shown in Figure 2-53 (c), and save the transformed coordinate system.

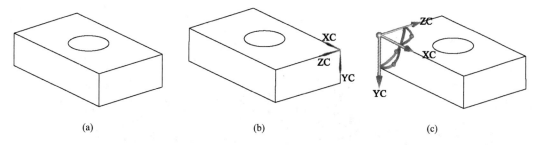

(a) (b) (c)

Figure 2-53　Layer, view, and coordinate system operations

Chapter 3 Sketch

A Sketch is a named set of 2D curves and points located on a specified plane or path.

You can apply rules in the form of geometric and dimensional constraints, to establish the criteria according to your design needs. Drawing a 2D sketch is the foundation of creating a 3D solid model. Features created from a sketch are associated with it if the sketch changes, so do the features.

You can use sketches to create:

(1) The profile or typical sections of your design;

(2) Detailed part features by sweeping, extruding, or revolving a sketch into a solid or a sheet body;

(3) Large-scale 2D concept layouts with hundreds, or even thousands of sketch curves;

(4) Construction geometry, such as a path of motion, or a clearance arc, that is not meant to define a part feature.

The main points of this chapter:

(1) Understand the NX design intent and modeling strategy;

(2) Introduction with creating sketch plane;

(3) Introduction with creating sketch curves;

(4) Sketch constraints;

(5) Sketch common actions.

3.1 Sketch Overview

3.1.1 The Basic Concept of Sketch

1. The sketch process

Typical steps involved in creating sketches as follows:

(1) Select a sketch plane or path.

(2) Choose your constraint recognition and creation options.

(3) Create the sketch geometry. Depending on your settings, the sketch creates many constraints automatically.

(4) Add, modify or delete constraints.

(5) Modify dimension parameters to meet your design intent.

(6) Finish the sketch.

2. Direct Sketch and the Sketch Task Bar

The *Direct Sketch* group and the *Sketch Task Bar* offer two modes you can use to create and edit sketches.

1) Direct Sketch

To create a sketch using *Direct Sketch* commands, click the *Sketch* command on the *Home tab* >*Direct Sketch* group in Modeling and other applications (Figure 3-1).

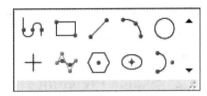

Figure 3-1   Toolbar of *Direct Sketch*

You can use the Direct Sketch group when you want to create or edit a sketch in the Modeling, Shape Studio, or Sheet Metal applications; or see the effect of sketch changes on the model in real-time.

2) Sketch Task Environment

The commands in the *Sketch Task Environment* are organized into multiple groups on the *Home* tab. This makes advanced commands more accessible on the Ribbon bar (Figure 3-2).

Figure 3-2   Toolbar of *Sketch Task Environment*

The users can access the *Sketch Task Environment* using one of these methods:

(1) Select the command *Sketch in Task Environment* on the *Curve* tab.

(2) From a feature creation dialog box, click the *Sketch Section* button. This creates a sketch internal to the feature using the *Sketch Task Environment*.

(3) When editing a sketch using *Direct Sketch*, select the command *Open* in *Sketch* in *Task Environment*.

(4) Right-click an existing sketch and choose *Edit with Rollback* to edit the sketch using the *Sketch Task Environment*.

Use the *Sketch Task Environment* when you want to:

(1) Edit an internal sketch.

(2) Experiment with sketch changes, but retain the option to discard all the changes.

(3) Create a sketch in other applications.

3. Sketch Constraints

In a sketch, you can fully capture your design intent through geometric and dimensional relationships as constraints.

Constraints can be used to create parameter-driven designs that you can update easily and predictably. And NX evaluates constraints as you work to ensure that they are complete and do not conflict.

While it is not required, it is recommended that you fully constrain sketches that define feature profiles.

In a sketch, you can create constraints as your design requires. That means you can use a sketch to create wireframe drawings that can serve a wide variety of up-front design purposes and are not meant for down-stream processing. For example you might create 2D layout sketches for products where you focus on product structure, component layout or basic component shape.

3.1.2 Sketch Preference

In the working environment of sketch, in order to draw the sketch more accurately and effectively, some general parameters need to be set before entering the sketch environment, so as to meet the different users.

The *Sketch Preferences* dialog box is called through the *Menu>Preference>Sketch Preference*. There are three tabs: *Sketch Settings*, *Session Settings* and *Part Settings*. The options are briefly described as below.

1. *Sketch Settings* tab

This tab can set basic parameters such as dimension label (expression, name and value) and text height of sketch, select the functions of creating inferred constrain and continuous auto dimensioning. The dialog box of the *Sketch Settings* tab is shown in Figure 3-3.

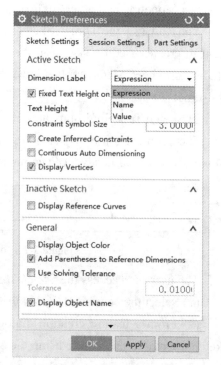

Figure 3-3 The dialog box of *Sketch Settings* tab

2. *Session Settings* tab

This tab can set basic parameters such as snap angle, sketch display state, and name prefix style in sketch, mainly including two panels of *Settings* and *Name Prefixes*. The dialog box of the *Session Settings* tab is shown in Figure 3-4.

3. *Parts Settings* tab

This tab allows you to use *Part Settings* to control color settings for curves, dimensions, degree-of-freedom arrows, and other sketch objects in the current part. *Part Settings* provides many of the same color options as *Sketch Colors* in *Customer Defaults*. However, when you modify colors in *Part Settings*, NX immediately applies your changes to all sketches in the current part.

Mastering the color of the sketch object is very useful for discovering timely possible problems or errors in the sketch. The dialog box of the *Part Settings* tab is shown in figure 3-5.

Figure 3-4 The dialog box of *Session Settings* tab   Figure 3-5 The dialog box of *Part Settings* tab

3.1.3 Create Sketch Plane

There are two types of sketch plane, one is *On Plane* and another is *On Path*.

1. Sketch on Plane

Sketch on an existing planar face, or on a new or existing sketch plane(Figure 3-6, Table 3-1). Key considerations that will guide your selection are:

Does the sketch define the base feature for the part? If so, create the sketch on an appropriate datum plane or datum coordinate system.

Is the sketch adding to an existing base feature? If so, select an existing datum plane or part face, or create a new datum plane with an appropriate relationship to existing datum planes or part geometry.

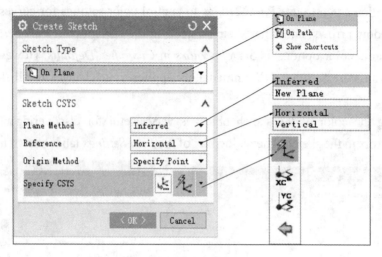

Figure 3-6　Dialog box of *Sketch on Plane*

2. Sketch on Path

You can use *Sketch on Path* to position a sketch for features like extrude and revolve. For all commands, you select the target path and define a sketch plane location on that path(Figure 3-7).

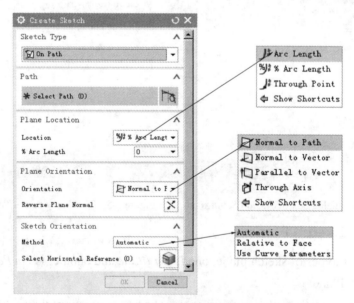

Figure 3-7　Dialog box of *Sketch on Path*

3.2　Sketching

3.2.1　Sketch Point

Point is widely used in NX. It is the construction of various curves, curved surface, solid, datum and other basic geometric elements. In the simple models, it is generally not necessary to set up points separately. In general, when spatial location cannot be determined by other ways, the creation of points is the most effective way.

You can click *Menu >Insert> Datum/Point >Point*, or you can click the shortcut toolbar ⊞ to open it directly, as shown in Figure 3-8.

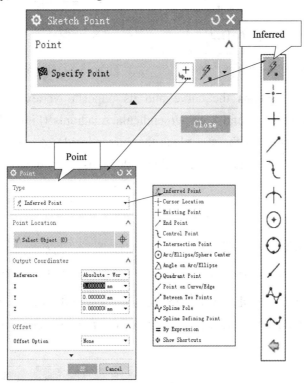

Figure 3-8 Dialog box of *Point*

The point is a single geometric object, using the *Point* dialog box, one is to select the point type to create the point by mouse capture, and the other is to directly input the coordinate position. The point can be further positioned by means of bias during both of the creation processes.

3.2.2 Profile

Use the *Profile* command ⌐ to create a series of connected lines and/or arcs in string mode. In string mode, the end of the last curve becomes the beginning of the next curve.

Click the *Profile* button on the *Sketch* toolbar and the *Profile* command dialog box will pop up, as shown in Figure 3-9.

When you transit from a line to an arc or from one arc to another arc, the quadrant zone symbol displays as shown in Figure 3-10.

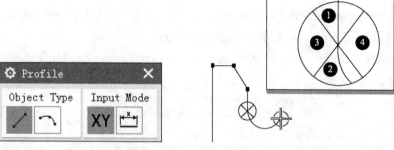

Figure 3-9 Dialog box of *Profile*   Figure 3-10 The arc quadrants

The quadrant that contains the curve and its opposite vertex are tangent quadrants (quadrants 1 and 2). Quadrants 3 and 4 are perpendicular quadrants (Figure 3-11, Figure 3-12).

To control the direction of the arc, place the cursor inside of one of the quadrants and then move the cursor out of the quadrant in either a clockwise or counterclockwise direction.

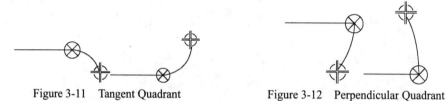

Figure 3-11 Tangent Quadrant   Figure 3-12 Perpendicular Quadrant

If the connection between the arc and the previous curve does not conform to the design intention, the movable cursor resets the end point of the previous curve and moves out from the appropriate quadrant.

3.2.3 Rectangle

Use the *Rectangle* command to create a rectangle using one of three methods: *By 2 Points*, *By 3 Points* and *From Center*.

By 2 Points creates a rectangle from 2 points in diagonal corners. The rectangle is parallel to the XC and YC sketch axes as shown in Figure 3-13.

By 3 Points creates a rectangle from a start point and two points that determine width, height, and angle. The rectangle can be at any angle to the XC and YC axes (Figure 3-14).

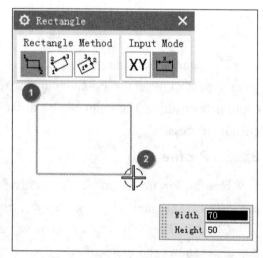

Figure 3-13 Rectangle *by 2 Points*

From Center creates a rectangle from a center point, a second point that determines angle and width, and a third point that determines height. The rectangle can be at any angle to the XC and YC axes (Figure 3-15).

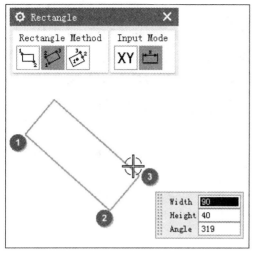
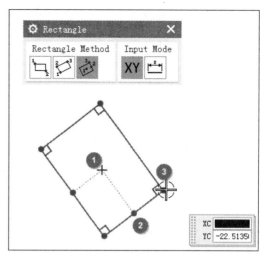

Figure 3-14 Rectangle by 3 Points   Figure 3-15 Rectangle from Center

3.2.4 Arc

Use the Arc command  to create an arc with either of two methods: *Arc by 3 Points* or *Arc by Center and Endpoints*.

Arc by 3 Points specify the arc start point 1 and end point 2, the point 3 decide the radius as shown in Figure 3-16.

Arc by Center and Endpoints specifies the arc center point and the start point and end point. The distance between the point 1 and the point 2 is the radius of the arc, point 2 and point 3 decide the sweep angle (Figure 3-17).

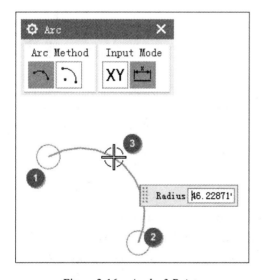
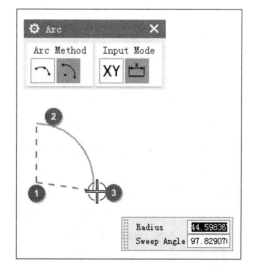

Figure 3-16 *Arc by 3 Points*   Figure 3-17 *Arc by Center and Endpoints.*

3.2.5 Circle

Use the *Circle* command to create circles with one of two methods:

Circle by Center or Diameter specify a center point and the diameter. The value of the diameter can be input in the dialog box or move the position of the point 2 (Figure 3-18).

Circle by 3 Points specifies two points on the circle, and a diameter. The value of the diameter can be input in the dialog box or move the position of the point 3 (Figure 3-19).

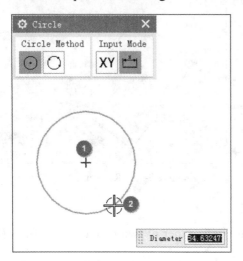
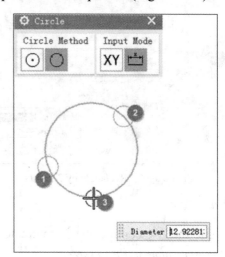

Figure 3-18　*Circle by Center and Diameter*　　　Figure 3-19　*Circle by 3 Points*

You can use either coordinate values or parameters for both methods. You can enter the diameter before you select the center point by selecting *Parameter Mode* (Figure 3-20).

3.2.6　Studio Spline

Use the *Studio Spline* command to dynamically create splines using either points or poles. The options are the same as the *Modeling Studio Spline* command, except sketch constraints can be used to constrain

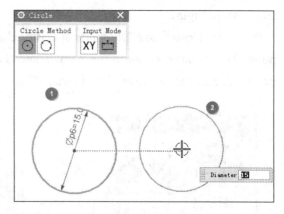

Figure 3-20　Circle stamping

pole locations and end tangency. There are two ways to create studio spline, through points and by poles (Figure 3-21, Figure 3-22).

Use snap options on the *Top Border* bar to snap to poles. For example, use the *Spline Pole* option to create dimensional constraints to pole locations. In commands that allow you to create or select a point, you can use the *Spline Pole* option in the point list to select spline poles.

While creating a spline, you can switch back and forth between the *Through Points* and *By Poles* options.

When switching from *Through Points* to *By Poles*, the through points and any internal constraints are deleted. Only start and end constraints are kept.

The spline cannot have fewer poles than the degree.

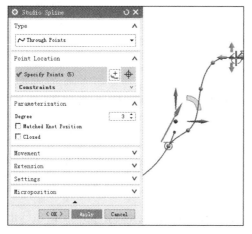
Figure 3-21　*Studio Splines Through Points*

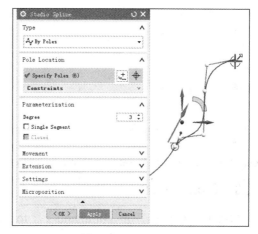
Figure 3-22　*Studio Splines By Poles*

3.2.7　Derived Lines

Use the *Derived Lines* command to create new lines from existing lines. You can create any number of offset lines from a base line, or a line midway between parallel lines, or a bisector line between non-parallel lines using this command.

(1) To offset a line from a base line, click the base, and click again to place the new line (Figure 3-23).

(2) To offset multiple lines from the same base line, hold <Ctrl> and click the base line. Then click again to place a new line (Figure 3-24).

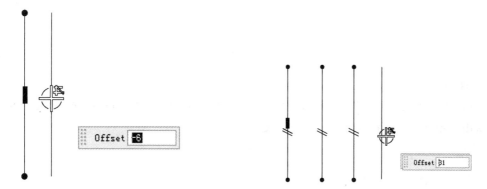

Figure 3-23　To offset lines from a base line　　　Figure 3-24　To offset multiple lines from the same base line

(3) To create a line at the intervening midpoint, select two parallel lines. You can set the line length by dragging the mouse or by entering a value in the *Length* input box(Figure 3-25).

(4) To construct a bisector line, select two nonparallel lines. You can place the line end graphically or enter a value in the *Length* input box. You can also place the bisector in any quadrant of the angled lines (Figure 3-26).

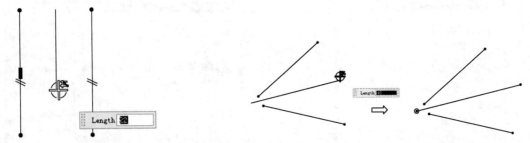

Figure 3-25 To create a line at the intervening midpoint   Figure 3-26 Construct a bisector line

3.2.8 Quick Trim

Use the *Quick Trim* command to trim a curve to the closest physical or virtual intersection in either direction. You can preview the trim by passing the cursor over the curve.

1. Trim individual curves

To trim individual curves, you can select individual curves to trim (Figure 3-27).

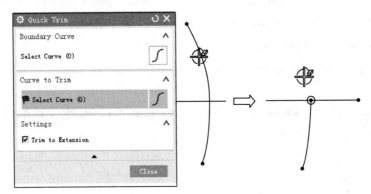

Figure 3-27 Trim individual curves

2. Trim multiple curves

To trim multiple curves, you can hold the left mouse button and drag across multiple curves to trim them all at the same time (Figure 3-28).

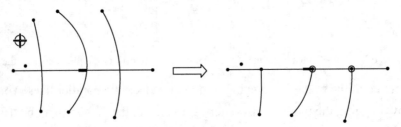

Figure 3-28 Trim multiple curves

Trimming a curve that has no intersection deletes the curve. A recipe curve is a curve projected to the sketch with associativity. If you delete a recipe curve by trimming, an alert message warns you that the associativity will be removed.

3.2.9 Quick Extend

Use the *Quick Extend* command to extend a curve to a physical or virtual intersection with another curve. You can preview the extension by passing the cursor over the curve.

1. Extend individual curves

You can select individual curves to extend (Figure 3-29).

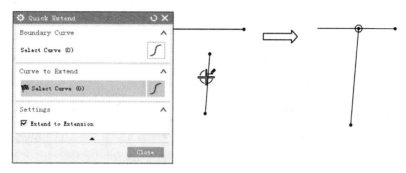

Figure 3-29   Extend individual curves

2. Extend multiple curves

You can hold the left mouse button and drag across multiple curves to extend them all at the same time (Figure 3-30).

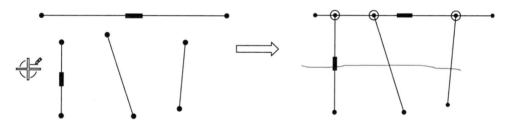

Figure 3-30   Extend multiple curves

3.2.10 Fillet

Use the *Fillet* command to create a fillet between two or three curves. You can trim all input curves or leave them untrimmed, or delete the third curve of a three-curve fillet, or specify a value for the fillet radius, or preview the fillet and determine its size and location by moving the cursor, or hold the left mouse button and drag over curves to create a fillet.

1. Fillet two curves

Choose *Home tab> Direct Sketch group> Fillet*. Then select the *Trim* or *Untrim* method as appropriate for your part. Then select the curves to fillet. Click left mouse button to complete the fillet.

You can move the cursor to adjust the fillet size and location. Change the value in the Radius input box. Press <Page Up> or <Page Down> to preview the complementary fillet (Figure 3-31).

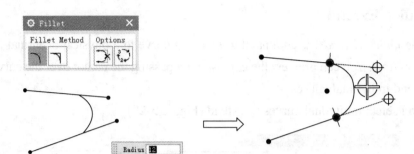

Figure 3-31　Fillet two curves

2. Fillet three curves

Choose *Home tab> Direct Sketch group> Fillet* . Select the Trim or Untrim method as appropriate for your part. Then select or clear *Delete Third Curve* as appropriate for your part. Then select curves one and two. To preview the fillet, move the cursor over the third curve. Click left mouse button to create the fillet.

If you delete the third curve, NX creates tangent constraints between the curves (Figure 3-32).

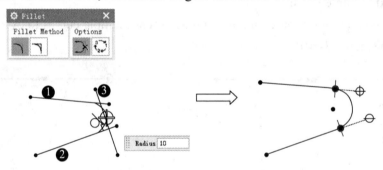

Figure 3-32　Fillet three curves

3.2.11　Chamfer

Use the *Chamfer* command to bevel a sharp corner between two sketch lines.

There are three types of chamfer offsets methods, *Symmetric*, *Asymmetric*, *Offset and Angle*. You can also hold the left mouse button and drag over curves to create a chamfer (Figure 3-33).

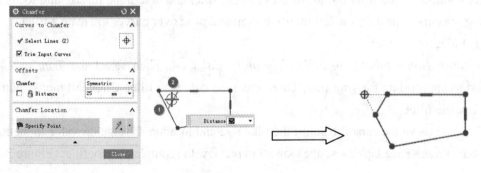

Figure 3-33　Chamfer

3.3 Constraints

Use *Constraints* to precisely control the objects in a sketch and to convey the design intent for a feature. There are two types of constraints, *Geometric Constraints* and *Dimensional Constraints*.

3.3.1 Basic Concept of Constraints

1. Geometric Constraints

Specify and maintain geometric conditions for or between sketch geometry. For example, *Geometric Constraints* can establish a line as vertical or horizontal, or two lines as perpendicular or parallel to each other.

2. Dimensions

Specify and maintain dimensions between sketch geometry. *Dimensional Constraints* are also called driving dimensions. For example, *Dimensions* can establish the size of a sketch object, such as the radius of an arc or length of a curve. And *Dimensions* also can create a relationship between two objects, such as the distance between two points.

3. Constraint conditions

If you enable the *Create Inferred Constraints* option, NX evaluates the sketch whenever you apply a constraint to identify. The constraint conditions are shown in Table 3-1.

Table 3-1 Constraint conditions

Constraint conditions	Description
Under constrained geometry	During constraint creation, NX displays degree-of-freedom arrows for curves or points that are under constrained. When you fully constrain a sketch: ① No degree-of-freedom arrows appear ② A Status message tells you that the sketch is fully constrained ③ The geometry changes to light green by default
Over constrained geometry	A curve or vertex is over constrained when you apply more constraints than are needed to control it. Geometry and any dimensions associated with it change to red by default
Conflicting constraints	Constraints can also conflict with each other. Conflicting constraints and the associated geometry in conflict change to magenta by default. NX displays the sketch in the last solved condition

Although you do not need to fully constrain a sketch to use it for downstream feature creation, the best practice is to fully constrain sketches. A fully constrained sketch ensures that a solution can be consistently found during design change. You can use a combination of *Automatic* and *Driving Dimensions*, and *Constraints* to fully constrain a sketch.

Whenever you encounter an over constrained or a conflicting constraint status, you should resolve the situation immediately by deleting some dimensions or constraints.

Perpendicular, Horizontal and Vertical dimensions maintain their direction when the expression value is set to zero. You can also enter negative values for these three dimension types to achieve the same results as using the *Alternate Solution* command. Avoid zero dimensions for other dimension types. Using zero dimensions leads to problems with ambiguous relative positioning to other curves. This can cause unexpected results when changing the dimension back to a nonzero value.

3.3.2 Geometric Constraints

Use the *Geometric Constraints* command to add geometric constraints to sketch geometry. These specify and maintain conditions for sketch geometry. You must first select a constraint type and then select the objects that you want to constrain. This workflow allows you to quickly create the same constraint on multiple objects.

Constraints can be used to define a line as being horizontal or vertical, to ensure that multiple lines remain parallel to each other, to require that several arcs have the same radius, or to position your sketch in space or relative to outside objects.

The example of *Geometric Constraints* is shown in Figure 3-34.

The *Geometric Constraints* dialog box is shown in Figure 3-35.

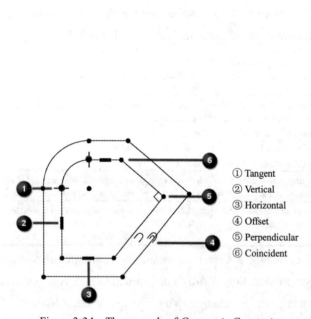

① Tangent
② Vertical
③ Horizontal
④ Offset
⑤ Perpendicular
⑥ Coincident

Figure 3-34 The example of *Geometric Constraints*   Figure 3-35 *Geometric Constraints* dialog box

When the *Select Object to Constrain* option is active, you can select multiple objects (Table 3-2). In the *Settings* group, you can select the:

Automatic Selection Progression check box so that you do not need to click the middle mouse button to advance to the *Select Object to Constrain* to option.

Constraints that you want to display by default in the *Constraints* group.

Table 3-2 The *Geometric Constraints* description

Type	Command icon	Description
Fixed		Defines fixed characteristics for geometry depending on the type of geometry
Fully Fixed		Creates sufficient fixed constraints to completely define the position and orientation of sketch geometry in one step
Coincident		Defines two or more points as having the same location
Concentric		Defines two or more circular and elliptical arcs as having the same center
Collinear		Defines two or more lines as lying on or passing through the same straight line
Point on Curve		Defines the location of a point as lying on a curve
Point on String		Defines the location of a point as lying on a projected curve. You must select the point first, then select the curve.
Midpoint		Defines the location of a point as equidistant from the two end points of a line or a circular arc
Horizontal		Defines a line as horizontal
Vertical		Defines a line as vertical
Parallel		Defines two or more lines or ellipses as being parallel to each other
Perpendicular		Defines two lines or ellipses as being perpendicular to each other
Tangent		Defines two objects as being tangent to each other
Equal Length		Defines two or more lines as having the same length
Equal Radius		Defines two or more arcs as having the same radius
Constant Length		Defines a line as having a fixed, unchanging length
Constant Angle		Defines a line as having a fixed, unchanging angle with respect to the sketch CSYS

3.3.3 Sketch Dimensions

Use *Sketch Dimensions* to establish the size of a sketch object, or the relationship between two objects in a sketch, or the relationship between two sketches, or the relationship between a sketch and another feature.

Sketch dimensions are displayed like drafting dimensions: they have dimension text, extension lines, and arrows. However, sketch dimensions differ from drafting dimensions because you can change the dimension value. This lets you control a feature derived from a sketch. Sketch dimensions also create an expression you can edit in the *Expressions* dialog box.

Perpendicular, *Horizontal*, *Vertical* and *Angular dimensions* maintain their direction when the expression value is set to zero. You can also enter negative values for these dimension types to achieve the same results as using the *Alternate Solution* command. The *Rapid Dimension* dialog box is shown in Figure 3-36.

There are five rapid dimension commands. Each command supports a related family of measurement methods. When you edit a dimension, you can change the measurement method between the families of measurement methods. The Dimension commands is shown Table 3-3.

Table 3-3 Dimension commands

Command	Icon	Description
Rapid Dimension		Creates a dimensional constraint between one or two objects you select. This command will infer one of these measurement types based on the objects you select, or you can explicitly select one these dimension measurement methods. A linear, radial, or angular dimension is created
Linear Dimension		Creates a dimensional constraint between the objects you select using one of these dimension measurement methods. When you edit a dimension using one of these measurement methods, you can change the measurement method between the methods listed
Radial Dimension		Creates a radial or diametral dimensional constraint on an arc or circle you select. When you edit a radial or diametral dimension, you can change the measurement method between the methods listed
Angular Dimension		Creates an angular dimensional constraint between two lines you select
Perimeter Dimension		Creates an expression to control the collective length of a set of lines and arcs you select. To view or edit the expression, use the *Expressions* or the *Edit Sketch Parameters* command

Figure 3-36 *Rapid Dimension* dialog box

3.3.4 Show/Remove Constraints

Use the *Show/Remove Constraints* command to display the geometric constraints that are associated with sketch geometry. You can use *Show/Remove Constraints* to remove specified geometric constraints, or to list information about all geometric constraints, or to interrogate and resolve over-constrained or conflicting conditions, or maintain design intent by checking for existing relationships to outside features or objects.

3.3.5 Convert To/From Reference

Use the *Convert To/From Reference* command to convert sketch curves from active to reference, or dimensions from driving to reference. Downstream commands do not use reference curves and reference dimensions do not control sketch geometry (Figure 3-37).

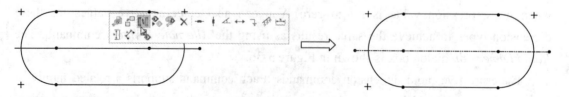

Figure 3-37 *Convert To Reference*

3.3.6 Alternate Solution

Use the *Alternate Solution* command to display alternate solutions for both dimensional and geometric constraints, and select a result. The example below shows how the geometry changes when you choose *Alternate Solution* and select a dimension (Figure 3-38).

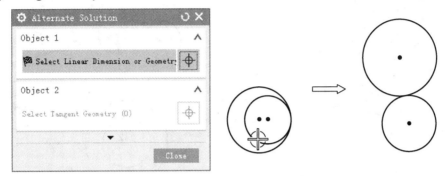

Figure 3-38 *Alternate Solution*

3.4 Exercise-Sketch Example

In this example, we will create the sketch of positioning plate as shown in Figure 3-37 to illustrate the details of sketching. After sketching successfully, we can build the 3D model of positioning plate. The main steps for sketching as follows:

(1) Choose *File* tab>*New*. Create a new Millimeters part based on the Model template.

(2) Enter *Dingweiban.prt* as the part name. Choose *Home* tab > *Direct Sketch* group> *Sketch*. Select the XY plane and then click the middle mouse button.

(3) Select the *Line* and *Arc* command, draw the lines and arcs according to Figure 3-39 respectively. Select the lines and arcs, right–click and choose *Convert To Reference*, the result as shown in Figure 3-40.

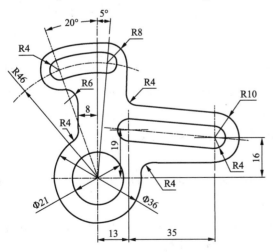

Figure 3-39 The sketch of positioning plate

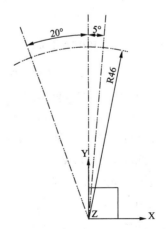

Figure 3-40 The reference lines

(4) Click the *Circle* button ⃝ and draw the circles according to the dimensions shown in Figure 3-39. After each circle is drawn, it needs to be completely fixed as shown in Figure 3-41.

(5) Click the *Line* button ╱ and draw a line above the two circles on the right side of Figure 3-41, as shown in Figure 3-42.

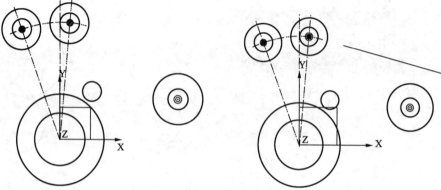

Figure 3-41　The circles　　　　Figure 3-42　The line

(6) Click the *Constraint* ⊥, select the line and the circle, click the *Tangent* button in the constraint operation panel . The result of adopting the tangential constraint is shown in Figure 3-43.

(7) Click *Quick Trim* to cut the unwanted line segments, as shown in Figure 3-44.

(8) Draw another line and make it tangent to the two circles in the same way, and use the *Quick Trim* to cut off unnecessary line segments, as shown in Figure 3-45.

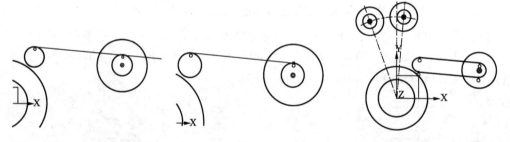

Figure 3-43　The tangent line　　　Figure 3-44　Trim　　　Figure 3-45　The another tangent line

(9) Click the *Line* ╱ to draw a straight line. Click the *Constraint* ⊥ to make it parallel to the created line and tangent to the right-most circle, as shown in Figure 3-46.

(10) Use the *Line* ╱ to draw a vertical line and use the constraint method to make it tangent to the middle circle, as shown in Figure 3-47.

(11) Click the *Fillet* , select the two edges that are tangent to it, and enter the fillet radius 4. The results are shown in Figure 3-48.

(12) Click the *Line* ╱, draw a horizontal line that is tangent to the bottom of the right—most circle with the capture function, as shown in Figure 3-49.

(13) Click the *Fillet* , complete the fillet as shown in Figure 3-50.

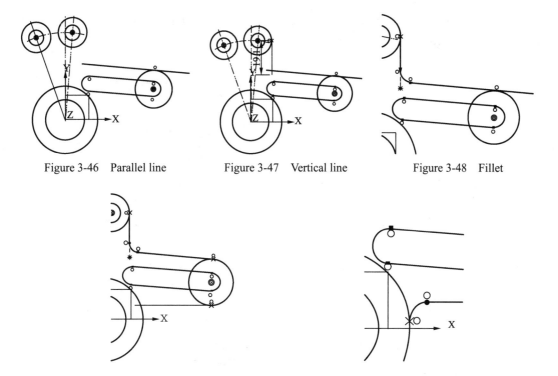

Figure 3-46 Parallel line Figure 3-47 Vertical line Figure 3-48 Fillet

Figure 3-49 Horizontal line Figure 3-50 Fillet vertical line

(14) Click *Quick Trim* to trim off the unnecessary lines on the right circle, as shown in Figure 3-51.

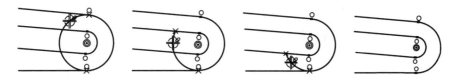

Figure 3-51 Trim

(15) Click the *Arc* and complete the two arcs with a sweep angle of 40 degrees according to the dimensions shown in the Figure 3-38, as shown in Figure 3-52.

(16) Click the *Quick Trim* to trim off unwanted arc sections, as shown in Figure 3-53.

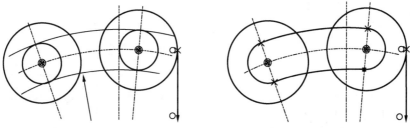

Figure 3-52 Arcs Figure 3-53 Trim arcs

(17) Click the *Arc* to draw two arcs that are tangent to the larger circle. Click *Quick Trim* to trim off the unnecessary lines on the right circle, as shown in Figure 3-54.

(18) Click the *Line* and draw a straight line parallel to the Y-axis with a distance of 8 on the left side, as shown in Figure 3-55.

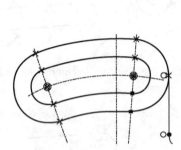

Figure 3-54 Tangent arcs

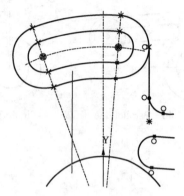

Figure 3-55 Straight line

(19) Click the *Fillet*, complete the rounded corners as shown in Figure 3-56.

(20) Click *Quick Trim* to trim off the unnecessary lines, as shown in Figure 3-57.

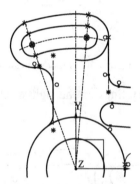

Figure 3-56 Fillet

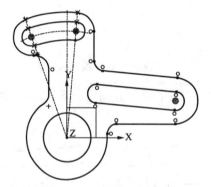

Figure 3-57 Trim

(21) Hide the reference lines, as shown in Figure 3-58.

(22) Click *Finish Sketch* to exit the sketch environment. A sketch can only be considered successful if it can be extruded. The 3D model of positioning plate is shown in Figure 3-59.

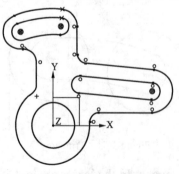

Figure 3-58 Finished sketch

Figure 3-59 The model of positioning plate

3.5 Questions and Exercises

1. Fill in the blanks

(1) _____ is a named set of 2D curves and points located on a specified plane or path.

(2) In the process of sketch drawing, _____ is the smallest geometric construction element, and also the basic element of sketch geometry elements.

(3) Use _____ to precisely control the objects in a sketch and to convey the design intent for a feature

2. Choice

(1) A _____ tool can be used to project the edge of a two-dimensional curve, solid or sheet along a direction onto an existing surface, plane or reference plane.

  A. Mirror Curve   B .Project Curve

  C. Offset Curve   D. Add Existing Curve

(2) _____ can draw a curve chain, and then trim all the curves that intersect the curve chain. The tool can be used to quickly trim multiple curves at a time.

  A. Single Trim   B. Uniform Trim   C. Border Trim   D. Delete

(3) _____ is used to control the dimensions of a sketch object or the relationship between two objects, which is equivalent to dimensioning the sketch.

  A. Auto Dimension   B. Dimensional Constraint

  C. Display/Remove Constraint   D. Continuous Auto Dimensioning

(4) _____ needs to specify the extension boundary, and the extended curve will extend to the boundary.

  A. Separate Extend   B. Uniform Extend

  C . Boundary Extend   D .Drag

3. Short answer questions

(1) Which methods are to create the working plane of sketch?

(2) Which ways are to draw a circle?

(3) Describe the geometric constraint mode of sketch?

4. Exercises

Please complete the following exercises for sketching training as shown in Figure 3-60. Please demonstrate the geometric elements, constraints and edit operation used during your sketching.

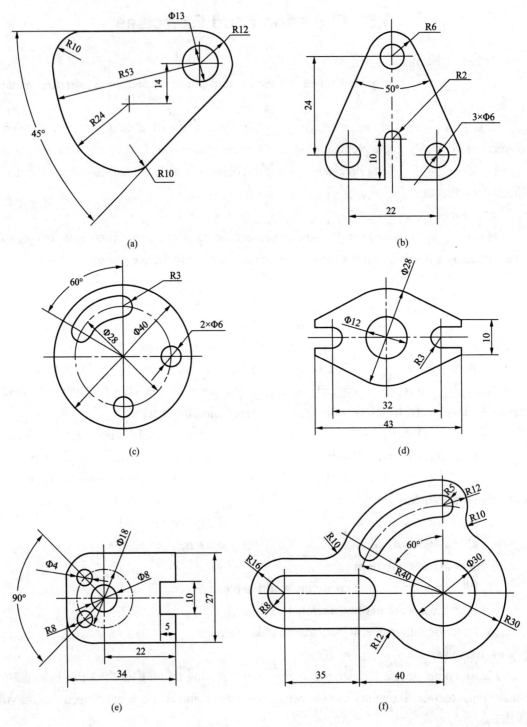

Figure 3-60　The sketch exercises

Chapter 4

Part Modeling

CAD module is the most fundamental module of 3D software. In this module, the most commonly modeling method is solid modeling, especially feature-based parametric solid modeling, and feature-based parametric solid modeling are basis and core modeling tool of CAD applications. In addition, users should be proficient in the type and creation method of features.

This chapter mainly introduces Boolean operations, basic features, engineering features, datum features and feature editing in NX.

After completing this chapter, the objectives to be achieved are:

(1) Understand the definition and type of features;

(2) Understand the characteristics of Boolean operations and principals of their operating methods;

(3) Become familiar with common design features so as to effectively use these tools for design.

4.1　Feature Modeling Overview

NX is one of the commonly used 3D CAD software. It employs feature-based parametric modeling techniques to support engineering designs through interactive user interfaces.

4.1.1　Common modeling methods

1. Parametric modeling

In the modeling phase of product design process, the designers should carefully analyze the structure of the product, understand the connection relationship of various parts of the product in order to form a reasonable design intent, and then use the functions provided by

CAD software to represent the physical design. The purpose of parametric modeling is to create solid models by rebuilding model as intended through varying its parameters. The parameters of the model can be related to each other to set up the connection relationship between the different components of the model. As shown in Figure 4-1, the parameters of the hole include the position, diameter and depth of the hole, and the parameters of the block include length, width and height. The designer's intent is that the hole stays at the middle of the block and the diameter of the hole is equal to half the width of the block, and the depth of the hole is constantly equal to the height of the block when the shape of block changes. To capture this design intent, the dimensions of the hole should be related to corresponding dimension of block by particular formula. By correlating these parameters, you can get the desired model that can be modified easily.

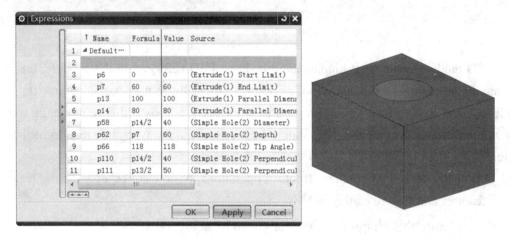

Figure 4-1　Parametric model

2. Constraint-based modeling

In constraint-based modeling approach, the position and structure of a model is driven or solved by using a set of plan policies referred to as constraints. These constraints can be dimension constraints (such as sketch dimension or positioning dimension) or geometric constraints (for example parallel or tangent, and so on). For instance, if a line is tangent to an arc, the design intent is to remain a tangent relation between the line and the arc when the angle of the line changes, as shown in Figure 4-2.

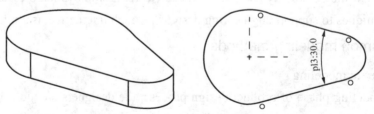

Figure 4-2　Model and its sketch

3. Feature-based Modeling

In 3D CAD software, a part model is usually composed of a large number of individual geometric elements with certain parameters, just as an assembly composed of many individual parts. These individual geometric elements, precisely called geometric features, contain both shape information and parametric information of a region interest. Features can be generated by methods such as sketch profile construction, feature tool and associative copy methods. These methods are usually integrated into modeling environment in NX.

4.1.2 Feature-based modeling process

The three-dimensional model of the part is composed of a series of features constructed in certain sequences (Figure 4-3). If the created features have a reference relationship between them, an association is established between them then a parent-child relationship is created between these features. The sequence of the parent feature and its child feature cannot be adjusted at will. In NX, this construction sequence is represented by a Timestamp order namely:

$$3D \text{ part model} = \Sigma \text{ Feature (Timestamp)}$$

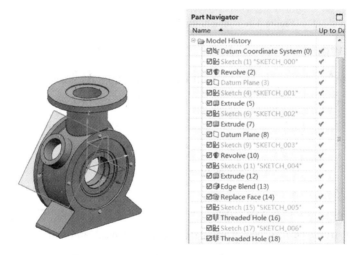

Figure 4-3　Composition of the box model

We can create parts with features by using methods such as the "building block", "revolving", and "machining" method. The building block method is to create only one simple feature at a time, and the latter features are then superimposed on the previous features as shown in Figure 4-4.

Figure 4-4　"Building Blocks" method to create a part

The revolving method is to create a section sketch profile for the part with the idea of mechanical drawing, and then revolve this sketch about an axis to obtain the part, which can quickly form the part but also limits the flexibility of subsequent model editing and modification as shown in Figure 4-5.

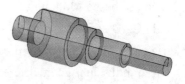

Figure 4-5 "revolving" method to create a part

The machining method is to follow the order of the parts in the manufacturing process and cut off the materials to create the part as shown in Figure 4-6. The sequences of this part should be fully understood during the creation process, which will affect the modeling efficiency.

Figure 4-6 "machining" method to create a part

4.1.3 Feature type

In 3D CAD software, some basic features, engineering features and datum features can usually be classified according to their roles in the modeling process. The basic features and engineering features belong to the forming features and the datum features are mainly used to provide references for forming features creation. Users can apply these features to create part models through performing related operations.

1. Basic features

The basic features are necessary for part solid modeling. Subsequent engineering features and feature editing must be based on the basic features. Basic features include design features such as extrude, revolve and sweep as shown in Figure 4-7, which are often used as the first feature when creating a part model.

2. Engineering features

Engineering features are solid modeling concepts introduced from engineering practice and further processing of basic features. Unlike the basic features, engineering features cannot be created independently and must be created on a face or an edge of existing objects. These features mainly include design features such as hole, boss, slot, groove, rib, thread, and detail features including chamfer, blend, and offset features such as shell as shown in Figure 4-8.

3. Datum features

The datum features mainly include features such as datum plane, datum axis, datum coordinate system, and datum point, as shown in Figure 4-9. There are two ways to create datum features: one is to create directly, and the other is to temporarily create during feature operation.

Figure 4-7   Basic features

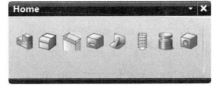
Figure 4-8   Engineering features

Figure 4-9   Datum features

4.2   Boolean Operations

Boolean operations are often used during NX modeling process. Boolean operations can be used to create features such as holes, bosses. In addition, they should be specified when creating basic features. Boolean operations are used to make more complicated objects by combining the existing objects in the model. When creating a new feature, Boolean operation also needs to be carried out on this new feature with existing objects so that they can be combined to form a single model. We can select an operation from the Boolean operation drop-down list box of the feature creation dialog box as shown in Figure 4-10 or select the corresponding operation from the menu *Insert > Combine*.

When performing Boolean operations, the objects that are being operated are called target body and tool body, respectively, according to their effect. The target body is the first body that needs to be merged with other bodies; the tool body is the body or object used to modify the target body. After completing the Boolean operation, the tool body will become part of the target body. The new creating feature is usually used as a tool body, and the existed features are taken as the target body.

In NX, three methods of *Boolean* operations are provided, namely unite, subtract and intersect. These are mathematical operations taken from set theory. As shown in Figure 4-11, an example model will be used to illustrate the *Boolean* operations and its options.

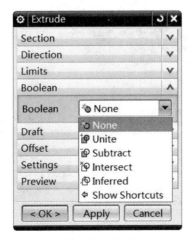
Figure 4-10   Choosing *Boolean* operations

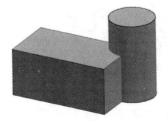
Figure 4-11   Example model

1. Unite

The unite operation refers merging of two or more bodies into single body. When performing the unite operation, the tool body and target body must overlap or share faces so that the result is a valid solid body. Click the *Unite* icon on the feature toolbar or select the menu *Insert > Combine > Unite* command to bring up the *Unite* dialog box. Then select the cuboid as the target body, and the cylinder as the tool body in turn, as shown in Figure 4-12. The target body and tool body are the same in following operations. The unite operation result is shown in Figure 4-13.

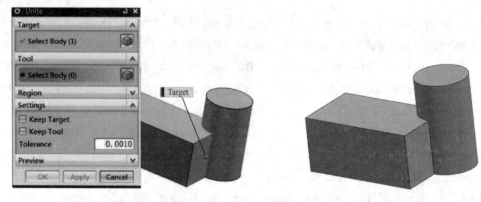

Figure 4-12   Unite operation   Figure 4-13   Unite operations result

A copy of the target body or tool body in an unmodified state also can be saved by selecting the *Keep Target* or *Keep Tool* option during the unite operation.

1) Keep Target

Selecting *Keep Target* check box in the *Settings* option of the *Unite* dialog box, a copy of the previously selected target body will be saved when performing the unite operation. The result of Boolean operation can be viewed by moving different bodies to different layers or by using the *QuickPick* tool as shown in Figure 4-14 and Figure 4-15.

 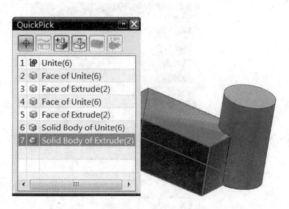

Figure 4-14   *Keep Target Setting* for unite operation   Figure 4-15   *Keep Target* operation result

2) Keep Tool

This check box in the *Settings* option of the *Unite* dialog box will result in saving a copy of the previously selected tool body when performing the unite operation. We can view the results of this option by moving different features to different layers or by using the *QuickPick* tool as shown in Figure 4-16 and Figure 4-17.

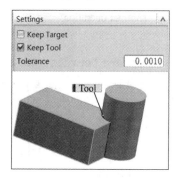 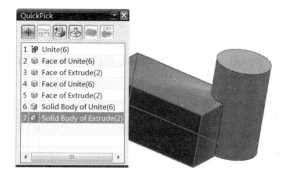

Figure 4-16　*Keep Tool Setting* for unite operation　　　Figure 4-17　*QuickPick* operation result

2. Subtract

The subtract operation is the removal of the volume of one or more tool bodies from a target body. The removed volume includes not only the specified tool bodies but also the intersection of the target body and the tool body. When performing the subtract operation, it is generally required that the tool and the target body must intersect each other.

Click the *Subtract* icon on the feature toolbar or select the menu *Insert* > *Combine* > *Subtract* command to show the *Subtract* dialog box. Then select the target body and the tool body in turn and perform the subtract operation as shown in Figure 4-18. Its result is shown in Figure 4-19.

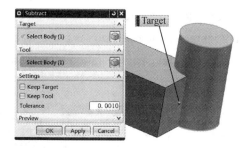 

Figure 4-18　Subtract operation　　　Figure 4-19　Subtract operation result

In the *Settings* option select the *Keep Target* check box. After performing the subtract operation, a copy of the target body is saved in the graphics area. When *Keep Tool* check box is only selected, a copy of the tool body is saved and still displayed in the graphics area after the subtract operation is performed. Their operation results are shown in Figure 4-20 and Figure 4-21, respectively.

Figure 4-20　Keep target setting result for subtract operation

Figure 4-21　Keep tool setting result for subtract operation

3. Intersect

The intersection operation can obtain the common volume or area between intersecting bodies. When performing the intersection, it is required that the tool body and the target body must intersect.

Click the *Intersect* icon on the feature toolbar or select the menu *Insert > Combine > Intersect* command to show the *Intersect* dialog box. Then select the target body and the tool body in turn and perform the intersection operation as shown in Figure 4-22. The result of intersect operation is shown in Figure 4-23.

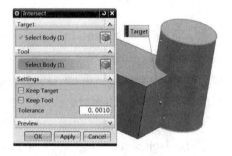   

Figure 4-22　Intersect operation

Figure 4-23　Intersect operation results

Select the *Keep Target* check box in the *Settings* option and a copy of the target body in an unmodified state will still be displayed in the graphics area after the intersection operation. Then *Keep Tool* check box is only selected and the intersection operation is performed, a copy of the specific tool body is still displayed in the graphics area. The operation effects are shown in Figure 4-24 and Figure 4-25.

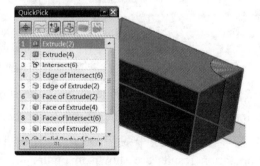   

Figure 4-24　Keep Target Setting result for intersect operation

Figure 4-25　Intersect operation

4.3 Basic Features

The basic features to design a part are extrude, revolve, sweep and many more. It is necessary to design its section, direction, limits, Boolean operations and other options to create this type of the feature.

4.3.1 Extrude Features

An extrude feature is a feature created by extruding a section a certain linear distance along a specified direction. Extruded sections can be two-dimensional geometric entities such as sketches, curves and so on. Click the *Extrude* button on the *Feature* toolbar to open the *Extrude* dialog box as shown in Figure 4-26. In this dialog box you can specify the parameters such as the section, direction, limits and Boolean operation of the extrude feature to create a solid or sheet body.

Figure 4-26　Extrude dialog box

1. Section

There are two ways to specify the section of the extrude feature by sketch section or selecting the existed curves as section. When creating a solid model, the section is often closed.

1) Sketch Section

Click the *Sketch Section* icon to open a sketch task environment. The user sketch a two dimensional (2D) profile as needed section for creating extrude feature. After the section is drawn, return to the *Extrude* dialog box.

2) Curve

Click the *Curve* to select curves, edges, a sketch, or a face for the section to extrude in the current environment. The user can quickly select the relevant entities by selection tool. Each complete curve or solid edge can constitute an extruded section.

2. Direction

Used to define the direction in which the section will be extruded to generate extrude feature. The default direction is the direction of normal line of the section plane. The user can also specify the extrude reference vector, that is *Specify Vector*, by selecting a curve or a vector created by method supported by that type. If the direction of the specified vector is exactly the opposite of the request then we can change its direction to the opposite side of the section by selecting *Reverse Direction* or double-clicking the arrow indicating the direction in graphics area.

3. Limits

Used to specify the limits of the extruded feature, as measured from the section. The *Start* option is used to specify the starting position of the extruded feature in the extruding direction and the *End* option is used to specify the end position of the extruded feature in the extruding direction.

1) Value

The limits of the extrude feature is determined by the distance entered by the user. When the input value is negative, it indicates that the start or end position of the limit is opposite to the current original direction as shown in Figure 4-27. The direction of the specified vector in the direction indicated by the arrow as shown in the figure 4-27. The dot handle indicates the starting position while the arrow handle indicates the ending position and the user can also set the corresponding distance by dragging these handles. The distance of the start and end is the distance of the handle relative to the plane of the extrude section.

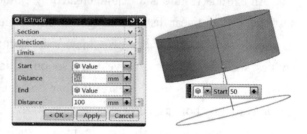

Figure 4-27   Value option settings

2) Symmetric Value

Converts the *Start* limit distance to the same value as the *End* limit. It means that the depth of the extrude feature is twice the distance value entered by the user, as shown in Figure 4-28.

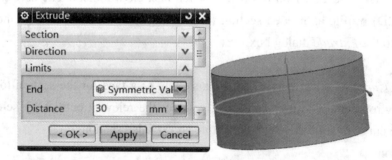

Figure 4-28   Symmetric Value

3) Until Next

The limits of the extruded feature is determined by finding an intersection with the next face that completely intersects the section in the model as shown in Figure 4-29. If there is no face in the direction of extrude that completely intersects the section, an appropriate error message will show.

4) Until Selected

The limits of the feature is determined by extending the section to a selected face, datum plane or body as shown in figure 4-30. It should be noted that the extrude section must extend past or completely intersect the selected reference object otherwise the extrude will fail.

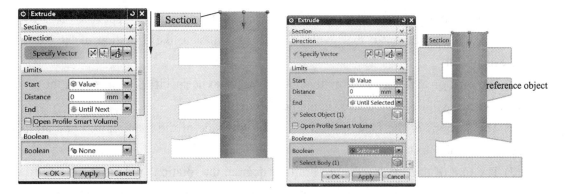

Figure 4-29 Until Next   Figure 4-30 Until Selected object

5) Until Extended

The limits of the extrude feature extends from the sketch plane to the selected reference face as shown in Figure 4-31. It should be noted that if the extrude section extends past or does not completely intersect the selected reference object in the extrude direction, the selected face will be extended so that it completely intersects the extended extrude section.

6) Through All

The limits of the extrude feature is from the sketch plane and extends to all selectable bodies along the path of the extrude direction as shown in Figure 4-32.

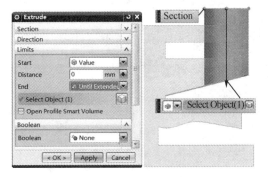 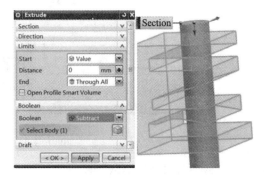

Figure 4-31 Until Extended   Figure 4-32 Through All

4. Boolean Operation

Used to specify that how the extrude feature interacts with bodies it comes in contact with when the extrude feature is generating. The *Boolean* option includes four options are *None*, *Unite*, *Subtract* and *Intersect*. *None* means no *Boolean* operation on the current feature, it is generally used for the first feature in the model. Usually, the software automatically determines the most probable Boolean operation according to the options such as the section, direction vector and limit of the objects being extruded, that is software automatically selects

Unite, *Subtract* or *Intersect* to perform Boolean operations.

4.3.2 Revolve features

A revolve feature is generated by rotating section curves around a specified axis by a certain angle. The operation of the revolve feature is similar to the operation of the extrude feature except that:

(1) The rotation axis vector of revolve feature needs to be specified. In addition, the point at that the vector will be located also should be specified;

(2) The sections of the feature must all lie on one side of the rotation axis and are not allowed to cross the axis;

(3) The unit of the revolve feature limits is degree; the rotation angle is calculated from the position of the section, and the sum of the absolute values of the start and end angles cannot be greater than 360°. Click the *Revolve* icon on the *Features* toolbar to open the *Revolve* dialog box, in which you can set the parameters of the revolve feature such as section, axis, limits and Boolean operation in order to create a round or partially round feature. Figure 4-33 shows the revolve feature created by rotating the YC axis of the datum coordinate system as the rotation axis vector and the datum coordinate origin as the axis vector position.

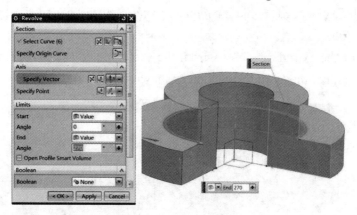

Figure 4-33   Revolve features

4.3.3 Sweep along Guide

A sweep along guide feature is to create a solid or sheet by sweeping an open or closed section consisting boundary sketch, curve, edge or face along a guide formed by one or a series of curves, edges, or faces. The section or guide can be open or closed. The section should be placed relative to the guide line. When the guide line is open, the section should start at the beginning of the guide line. When the guide line is closed, the section should start at the beginning of any curve making up the guide line. Sweeping along guide feature requires that the section cannot self-intersect during sweeping process as shown in Figure 4-34. Some following instructions are very useful to avoid unexpected errors:

(1) The guide line should be smooth, no sharp corners;

(2) The radius of curvature of curve on the guide line is larger than the radius of curvature of the corresponding portion on the section.

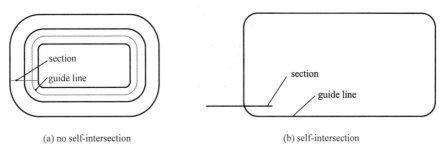

(a) no self-intersection   (b) self-intersection

Figure 4-34   Sweep feature when section is closed

From the pull-down menu, choose *Insert > Sweep > Sweep along Guide*, and you will see the *Sweep along Guide* dialog box in which user can complete the creation of the feature by setting options such as section, guide, offsets, and Boolean.

1) Closed guide line

When the guide line is closed and the section is closed as shown in Figure 4-35, the sweep feature is a solid volume formed by the closed section along the closed guide line as shown in Figure 4-36.

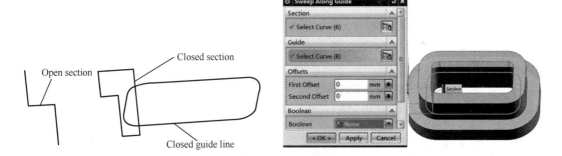

Figure 4-35   section and closed guide line   Figure 4-36   Sweep feature when section is closed

When the guide line is closed and the section is open as shown in Figure 4-36, then the sweep feature is solid volume formed by filling area surrounded by the open section sweeping along the closed guide line as shown in Figure 4-37.

2) Open guide line

When the guide line is open and the section is closed as shown in Figure 4-38, then the sweep feature is solid volume formed by the closed section sweeping along the open guide line as shown in Figure 4-39.

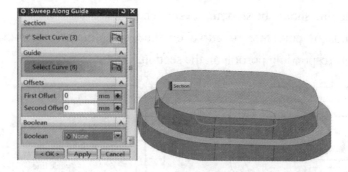

Figure 4-37 Sweep feature when the section is open

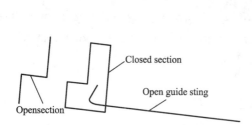

Figure 4-38 section and open guide line   Figure 4-39 Sweep feature with closed section

When the guide line is open and the section is open as shown in Figure 4-38, then the sweep feature is a sheet formed by the open section sweeping along the open guide line as shown in Figure 4-40. At this case, we can also generate a solid feature by setting the *Offsets* option as shown in figure 4-41.

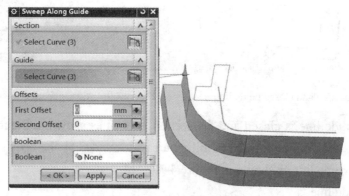

Figure 4-40 Sweep feature with open section

The first *Offset* option can offset the *Sweep* feature to add thickness, and the second *Offset* option offsets the base of the *Sweep* feature away from the section string.

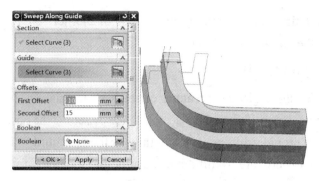

Figure 4-41 Sweep Feature with Offset option

4.4 Placement of Engineering Features

The engineering features, also called as pick-and-place features, mainly include holes, bosses, slots, grooves, ribs, threads, chamfer, fillets and shell. These features are often added in the final stage of the solid modeling process after the basic features are created. When placing such features on a face or an edge of existing objects, you need to specify their position relative to the existing feature in addition to specifying the corresponding shape parameters based on the feature type.

4.4.1 Placement Face

The placement face for engineering features is usually planar face. The placement face of the groove feature is exceptional and its placement face must be a cylindrical or conical face of solid body.

The placement face is usually the face on the existing solid. If no suitable face is available for the placement face then user can also use the datum plane as the placement face as shown in Figure 4-42. In order to place a hole on a cylindrical face, it is necessary to make a datum plane tangent to the cylindrical face as the placement face on which to place the hole.

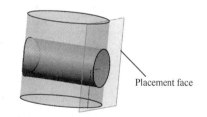

Figure 4-42 Datum plane as the placement face

4.4.2 Horizontal Reference

The horizontal reference is used to define the XC direction of the feature coordinate system. You can select an edge, face, datum axis, or datum plane as the horizontal reference(Figure 4-43).

A horizontal reference is usually required in the following cases:

(1) For engineering features with length parameters, in order to measure the length of these features, a horizontal reference needs to be specified to define the length direction.

(2) In order to define the positioning dimensions of the horizontal or vertical type, you also need to specify the horizontal reference.

4.4.3 Positioning methods

In order to position the engineering features correctly on the placement face, we also need to specify their positioning dimensions on the placement face by setting the positioning methods.

The system will provide different positioning methods according to different types of engineering features. There are 6 different positioning methods for circular or conical engineering features. For rectangular engineering features, there are 9 different positioning methods as shown in Figure 4-44. The meaning of each type of positioning method is shown in Table 4-1.

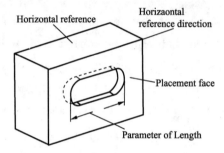

Figure 4-43   Horizontal reference of the Slot feature

Figure 4-44   *Positioning* dialog box for rectangular engineering features

Table 4-1   Positioning methods

Name	Icon	Meaning
Horizontal		Create a positioning dimension between two points. The horizontal dimension is aligned with the horizontal reference or 90° to the vertical reference
Vertical		Create a positioning dimension between two points. The vertical dimension is aligned to the vertical reference or 90 degrees to the horizontal reference
Parallel		Create a positioning dimension between two points. It is the shortest distance between two points
Perpendicular		Specify a linear edge, datum, or the shortest distance between the axis and a point
Parallelata istance		Specifies a linear edge on the tool feature that is parallel to a linear edge, datum or axis on the target feature at a given distance
Angular		Create a positioning dimension measured by angle between two linear edges
Point onto point		Specifies that the distance between two points is zero, even if the target object and the selected point on the tool object coincide
Point onto line		Specifies that the distance between a point and an edge, datum, or axis is zero, even if the point on the tool object is on the reference object of the target
Line onto line		Specifies that the distance between the two sides is zero. Even if the selected object and the selected edge on the tool object coincide

Positioning dimensions create an associativity between the feature and the target solid. All types of positioning dimensions require measurement between two selected points or two objects. The first point or the first object is the reference object on the existing entity and it is used as the target reference for the positioning dimension. The second point or the second object is the reference object on the engineering feature to be created and it is used as the reference object.

4.5 Datum Features

The datum features are auxiliary entities that aid solid model creation, including the datum plane, the datum axis, the datum coordinate system, and the datum point. The datum plane and the datum axis can be established relative to the existing solid model and the datum established at this time are called the relative datum. Default datum features, including a coordinate system and three perpendicular planes, are provided in software as starting point for part solid modeling (and assembly). As these datum features do not reference any other geometry, we called them as fixed datum feature.

4.5.1 Datum Plane

The *Datum Plane* command is used to create a planar reference feature to help define other feature. The relative datum features are established by referencing curves, edges, points, surfaces, or other datum features on the existing model. The fixed datum features do not reference other geometry entities. You can use either of the methods creating a relative datum to create a fixed datum by canceling the *Associative* check box in the *Datum Plane* dialog box. You can also create a fixed datum based on the WCS and the absolute coordinate system (ACS).

The datum plane can be used as sketch plane, a placement face for engineering features, a target reference for locating engineering features, a reference to basic feature limits, and constrained positioning in an assembly.

Usually in a model, multiple datum planes can be created, but generally only three coordinate planes, such as XC-YC, YC-ZC and YC-ZC, are established.

Choose *Insert > Datum/Point > Datum Plane* command or click the *Datum Plane* on the feature toolbar to show the Datum Plane dialog box as shown in Figure 4-45. Pull down the *Type* list to choose a datum plane creation method. Table 4-2 lists the methods for constructing a datum plane and their meanings.

 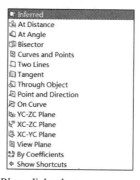

Figure 4-45 *Datum Plane* dialog box

Table 4-2 Methods for creating a datum plane

Name	Icon	Meaning
Inferred		Determines the best datum plane type to use based on objects you select
At Angle		Create a plane by specifying a planar object as a reference and specifying an angle with the plane
At Distance		Creates a plane parallel to a planar face or another datum plane at a distance you specify
Bisector		Creates a plane midway between two selected planar faces or planes. If the input planes are at an angle to each other, the plane is placed at the bisected angle
Curves and Points		Creates a plane using various combinations of points, a line, a planar edge, a datum axis, or a planar face (for example, three points, a point and a curve, etc.)
Two Lines		Creates a plane using a combination of any two linear curves, linear edges, or datum axes
Tangent		Creates a datum plane tangent to a non-planar surface, relative to a second selected object
Through Object		Creates a datum plane on the surface normal of a selected object
Point and Direction		Creates a plane from a point and a specified direction
On Curve		Creates a plane at a location on a curve or edge
By Coefficients		Create a fixed, non-associative datum plane on the WCS or absolute coordinate system using an equations of A, B, C, and D coefficients: $Ax + By + Cz = D$
YC-ZC Plane		Create a fixed datum plane along the YC-ZC plane of the Work Coordinate System (WCS) or Absolute Coordinate System (ACS).
XC-ZC Plane		Create a fixed datum plane along the XC-ZC plane of the Work Coordinate System (WCS) or Absolute Coordinate System (ACS).
XC-YC Plane		Create a fixed datum plane along the XC-YC plane of the Work Coordinate System (WCS) or Absolute Coordinate System (ACS)
View Plane		Create a fixed datum plane parallel to the view plane and passing through the origin of the WCS.

4.5.2 Datum Axis

The *Datum Axis* command is used to define the datum axis feature. The datum axis is divided into a relative datum axis and a fixed datum axis. The relative datum axis is associated with one or more datum objects, the fixed datum axis does not reference other geometry and it is non-associative. The datum axis can be used as an axis for the revolve feature, a reference to the circular array, a reference to the datum plane, and the like.

Choose *Insert > Datum/Point > Datum Axis* menu command or click the *Datum Axis* on the feature toolbar to bring up the *Datum Axis* dialog box as shown in Figure 4-46. Select a type in the list to create a datum axis. Table 4-3 lists the methods for creating a datum axis and their meanings.

Figure 4-46 *Datum Axis* dialog box

Table 4-3 Method of creating a datum axis

Name	Icon	Meaning
Inferred		Determine the best way to create a datum axis based on the object selected by the user
Intersection		Creates a datum at the intersection of two datum planes or planes
Curve/Face Axis		Create a datum axis along a linear or linear edge, or an axis of a cylindrical, conical, or toroid
On Curve Vector		Create a datum axis that is tangent to a specified point on a curve or edge, perpendicular or bidirectional, or perpendicular or parallel to another object
XC-axis		Create a fixed datum axis on the XC axis of the working coordinate system (WCS)
YC-axis		Create a fixed datum axis on the YC axis of the working coordinate system (WCS)
ZC-axis		Create a fixed datum axis on the ZC axis of the working coordinate system (WCS)
PointandDirection		Create a datum axis in the specified direction after the specified point
Two Points		Specify two points through which to create a datum axis

4.6 Exercises - Feature Modeling

This exercise illustrate step-by-step details for creating a solid part model using extrude features, sweeping features along the guide path, and revolve features as shown in Figure 4-47.

The steps are as follows:

(1) Open the exercise *ch4\example\4-1.prt*, as shown in Figure 4-48.

Figure 4-47 Feature Modeling Example Model   Figure 4-48 4-1.prt

(2) Create an extrude feature. Choose *Insert > Design Features > Extrude* command or click the *Extrude* button on the *Feature* toolbar to bring up the *Extrude* dialog box.

(3) Select the sketch as shown in Figure 4-48 as the extrude section, specify the ZC axis

as the extrude direction, and then set the limits value of the extrude. Click the middle mouse button to complete the extrude feature as shown in Figure 4-49.

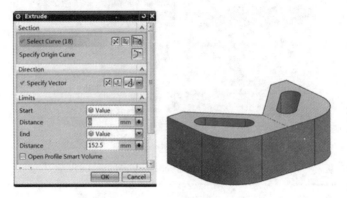

Figure 4-49   Extrude feature

(4) Create a sweep feature along the guide path. Select *Sketch (8)* and *Sketch (10)* in the *Part Navigator* to display them as shown in Figure 4-50. Choose *Insert > Sweep > Sweep along Guide* and open the *Sweep Along Guide* dialog box.

(5) Select *Sketch (10)* as the section curves, click the middle mouse button, select *sketch (8)* as the guide path. Select the *Unite Boolean* operation and select the extrude feature as the target body. Click the middle mouse button to complete the creation of the sweep feature as shown in Figure 4-51.

Figure 4-50   Sweep string   Figure 4-51   Sweep along guide

(6) Create a revolve feature. Hide the *sketch (8)* and the *Sketch (10)* and display the *sketch (6)* as shown in Figure 4-52. Click the *Revolve* button on the *Feature* toolbar to open the *Revolve* dialog box.

(7) Select *sketch (6)* as the section and select the vertical line of *sketch (6)* as the axis vector. Set the start limit to 0 degree and end limit to 360 degree. Select *Unite* Boolean operation then click the middle mouse button to complete the creation of the revolve feature as shown in Figure 4-53.

(8) Remove the volume by extrude feature. The *sketch (7)* is displayed, as shown in

Figure 4-54. Click the *Extrude* button on the Feature toolbar to bring up the *Extrude* dialog box.

Figure 4-52   Revolve feature sketch   Figure 4-53   Revolve featur   Figure 4-54   Sketch for extrude

(9) Select *sketch (7)* as the section; the direction of the extrude points to the existing entity. The start limit is 0 and the *End* option is *Until Next*. Note that in the *Boolean* group, *Subtract* is selected. Click the middle mouse button to complete the extrude as shown in Figure 4-55.

(10) Save model.

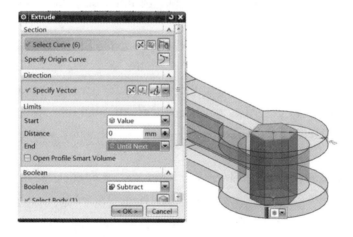

Figure 4-55   Extrude feature

4.7 Modeling of Typical Mechanical Parts

4.7.1 Modeling of gear shaft

Shaft parts are the most common mechanical parts, as shown in Figure 4-56. Shaft parts are mainly used to support transmission components such as gears, pulleys, cams and connecting rods to transmit torque. According to the function and structural shape, the shaft parts have various forms such as hollow shaft, half shaft, step-shaped shaft, spline shaft, crank shaft and gear shaft. The length of the shaft parts is larger than the diameter. The basic structure is a thin-solid-shaped or hollow rotary body and the surface is usually distributed with a center bore, a key slot, a groove and a chamfer, etc. Sleeve-type parts are simpler mainly used for distance and isolation.

(a) Stepped shaft　　　　　(b) geared shaft　　　　　(c) sleeve

Figure 4-56　Shaft part model

This section will create the gear shaft model shown in Figure 4-57. The part consists of gear, stepped shaft, chamfers, fillets and slot. The detailed modeling process for this model is as follows:

1. Create a new file

Figure 4-57　Gear shaft model

After starting NX, click the New button on the toolbar to bring up the *New* dialog box. In the *Model* tab, select the *Model* template, the unit is mm, and enter the file name *4-3.prt*. Click the middle mouse button to enter the *Modeling* module.

2. Create gear

NX provides a special gear toolbox, the *Gear Modeling* in *GC Toolkit* to create and edit common gear parts and assemblies.

The gear modeling tool provides the creation of two types of gears called as cylindrical gear and bevel gear. The bevel gears are divided into straight bevel gear and helical bevel gear. These gears are created in much the same way except for specific geometric parameters.

(1) Choose *GC Toolkits* > *Gear Modeling* > *Bevel Gear* or click the *Bevel Gear Modeling* button on the feature toolbar to show the *Bevel Gear Modeling* dialog box as shown in Figure 4-58. Select *Create Gear* as *Gear Operating Type* and click the middle mouse button to show the *Bevel Gear Type* dialog box as shown in Figure 4-59. Select the gear type as *Helical Gear* and the *Tooth Type* is *Equal Clearance Tooth*. Click the middle mouse button to show the *Bevel Gear Parameters* dialog box as shown in Figure 4-60.

 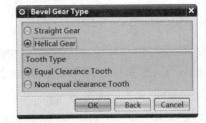

Figure 4-58　*Bevel Gear Modeling* dialog box　　　Figure 4-59　*Bevel Gear Type* dialog box

(2) Enter the gear parameters. In the dialog box as shown in Figure 4-60, click

Default Value, the software automatically assigns the gear name and parameters, or you can enter the corresponding parameters. Click *Parameter Estimation* to bring up the dialog box shown in Figure 4-61. Enter the corresponding parameters. The software will estimate the corresponding parameters to help the user complete the parameter input. After the user enters the correct parameters, click the middle mouse button to show the *Vector* dialog box.

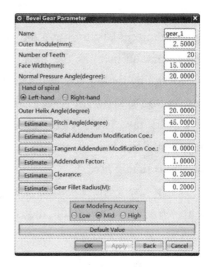

(3) The *Vector* dialog box is used to specify the axis direction of the gear. In the graphics area, select the Z-axis positive direction as the vector direction as shown in Figure 4-62. Click the middle mouse button to show the *Point* dialog box.

Figure 4-60 *Bevel Gear Parameter* dialog box

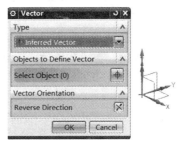

Figure 4-61 *Input Parameter of Match Gear* dialog box

Figure 4-62 *Vector* dialog box

(4) The *Point* dialog box is used to specify the vertex position of the gear. Select the coordinate origin as the vertex position in the graphics area as shown in Figure 4-63. Click the middle mouse button to complete the creation of the gear. The generated gear is shown in Figure 4-64.

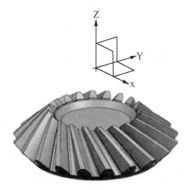

Figure 4-63 *Point* dialog box

Figure 4-64 generated gear model

(5) In order to coincide the position of the center of the small end of the gear with the origin, it is necessary to move the gear. Choose *GC Toolkits > Gear Modeling > Bevel Gear* menu command or click the *Bevel Gear Modeling* on the feature toolbar to show the *Bevel Gear* Modeling dialog box and select *Move Gear* as shown in Figure 4-65. Click the middle mouse button to show the *Select Gear to Operate* dialog box.

(6) In the *Select Gear Operation* dialog box, select the gear to be moved in components list box as shown in Figure 4-66. Click the middle mouse button to show the *Move Gear* dialog box.

(7) In the *Gear Position* dialog box, click the *Point to Point* button, as shown in Figure 4-67. Click the middle mouse button to show the *Point* dialog box.

Figure 4-65　Bevel Gear Modeling Dialog Box　　　　Figure 4-66　Select Gear Operation dialog box

(8) In the *Point* dialog box, select center of the small end and the coordinate origin in turn, as shown in Figure 4-68. The moved gear is shown in Figure 4-69.

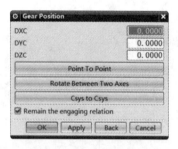

Figure 4-67　Gear Position Dialog Box　　　　Figure 4-68　Select Point

3. Create other shaft segments

(1) Choose from the pull-down menu command *Insert > Design Feature > Revolve* or click the *Revolve* on the feature toolbar to show the *Revolve* dialog box. In the *Section* option, click *Sketch Section* to bring up the *Create Sketch* dialog box.

(2) In the *Create Sketch* dialog box, select the XZ coordinate plane as the sketch plane, and the Z-axis positive direction as the horizontal direction of the sketch, as shown in

Figure 4-70. Click the middle mouse button to enter the sketch environment.

Figure 4-69　moved gear　　　　　　　Figure 4-70　Sketch option settings

(3) Sketch the section shown in Figure 4-71. Click the *Finish Sketch* button on the toolbar to return to the *Revolve* dialog box.

(4) In the *Axis* option, select the Z axis of the datum axis as *Specified Vector*, and the revolving angle is 360 degree. Click the middle mouse button to complete the creation of other shaft segments as shown in Figure 4-72.

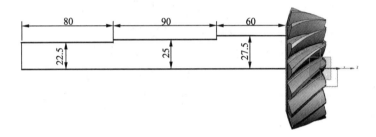

Figure 4-71　Sketch section

4. Create a slot

(1) Create a datum plane as the placement face for the slot. Select the outer cylindrical surface of the shaft as a reference to construct the datum plane as shown in Figure 4-73.

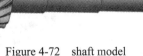

Figure 4-72　shaft model　　　　　　　Figure 4-73　creating a datum plane

(2) Select the menu command *Insert > Design Features > Slot* or click *Slot* on the feature toolbar to show the *Slot* dialog box. Select the *Rectangular* as slot type, click the middle mouse button to show the *Rectangular Slot* dialog box, select the datum plane

made in step 1 as the placement face, and click the middle mouse button to accept the default options. Select the datum Z axis as a horizontal reference to specify the direction of the slot length.

(3) In the *Edit Parameters* dialog box, set the slot parameters, as shown in Figure 4-74. Click the middle mouse button to show the *Location* dialog box.

(4) Set the positioning dimension in the horizontal direction to 33 and the positioning dimension in the vertical direction to 0, as shown in Figure 4-75.

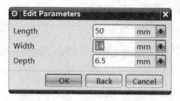

Figure 4-74   *Slot Edit* parameters   Figure 4-75   Setting the positioning dimensions

5. Create chamfers

(1) Select the menu command *Insert > Detail Feature > Chamfer* or click the *Chamfer* icon on the feature toolbar to show the *Chamfer* dialog box.

(2) The chamfer parameters setting is shown in Figure 4-76, in which the bottom edge of the end face of the shaft is selected as a reference for *Edge* option, the *Cross Section* and *Distance* in the *Offsets* options are set to *Symmetric* and 3, respectively. That is, the offset distance on each side of the selected edge is 3. Click the middle mouse button to complete the creation of the chamfer feature.

6. Create face Blend

(1) Select the menu command *Insert > Detail Feature > Face Blend* or click the *Face Blend* icon on the feature toolbar to show the *Face Blend* dialog box.

(2) Select the outer face of the big end of the gear and the outer cylindrical face of the shaft as the face chain 1 and the face chain 2, adjust the direction of the reference plane of the blend by the *Reverse* button; set the *Radius* to 3 as shown in Figure 4-77.

Click the middle mouse button to complete the creation of the face blend.

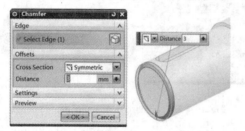
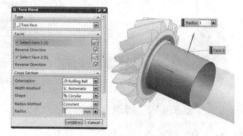

Figure 4-76   Chamfer Feature   Figure 4-77   Face Blend

7. Create a groove

(1) Select the menu command *Insert > Design Feature > Groove* or click the *Groove* on the

feature toolbar to show the *Groove* dialog box, as shown in Figure 4-78. This dialog provides three types of slot that are created in much the same way except for specific shape parameters.

(2) Select the *Rectangular* and click the middle mouse button to show the *Rectangular Groove* dialog box, select the cylindrical surface where the second key slot is located as the placement face; in the new *Rectangular Groove* dialog box, enter the groove parameters, as shown in the Figure. 4-79. Click the middle mouse button to show the *Positioning* dialog box.

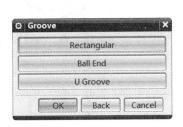

Figure 4-78   *Groove* dialog box   Figure 4-79   *Rectangular Groove* dialog box

(3) Select the edge shown in Figure 4-80 as the target entity and tool entity in turn, and enter the positioning dimension value as 0. Click the middle mouse button to accept the setting and complete creation of the slot feature, as shown in Figure 4-81.

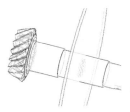   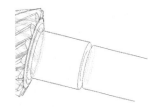

Figure 4-80   Positioning reference of the groove   Figure 4-81   Groove features

Finally save the file and close the current window.

4.7.2 Modeling of case part

The internal and external structures of the box-type parts are very complicated. Since it is mainly used to support and accommodate moving parts or other parts, they are often hollowed out to form cavity. The cavity of the case is usually used to install the drive shaft, the gear (or cam) and the rolling bearing. Therefore, there are holes for bearing caps and sleeves at both ends. And there are many mounting holes, positioning holes and connecting holes, etc. for assembling case. In order to tightly and precisely assemble the case, the case body is often provided with a flange. Since the inner section of case is a cavity, usually the wall is relatively thin. In order to increase the stiffness of the case body, such parts are generally provided with ribs for reinforcing its stiffness. Due to their complex shapes, they also have many casting process structures, such as casting fillets and draft angles as shown in Figure 4-82.

Create the case part shown in Figure 4-83. The part consists of features such as extrude, shell, holes, blend, and ribs. The detailed modeling process for this part is as follows:

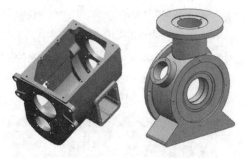

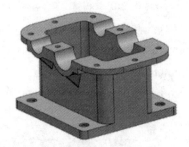

Figure 4-82   case parts

Figure 4-83   Case parts

1. Create a new file

After starting NX, click the New button on the toolbar to bring up the *New* dialog box. In the *Model* tab, select the Model template, the unit is mm, and enter file name *4-4.prt*. Click the middle mouse button to enter the modeling module.

2. Create a case base

(1) Select the menu command *Insert > Sketch*, choose the XY plane of datum coordinate system as the sketch plane in the graphics area, select the X axis positive direction as the horizontal reference of the sketch orientation, and click the middle mouse button to enter the sketch task environment.

(2) Create the sketch shown in Figure 4-84 and click the *Finish Sketch* on the toolbar to exit the sketch environment.

(3) Select the menu command *Insert > Design Feature > Extrude* or click *Extrude* on the feature toolbar to show the *Extrude* dialog box. In the *Section* option, select the sketch created in the previous step and select the Z axis positive direction as the extrusion direction. The setting of the *Limits* option is shown in Figure 4-85. Click the middle mouse button to complete the creation of the extruded feature.

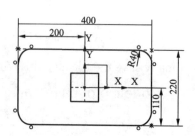

Figure 4-84   Sketch of the base section

(4) Click *Extrude* on the feature toolbar to bring up the *Extrude* dialog box. In the *Section* option, select the upper surface edge of extrude feature created in the previous step as the sketch for the extruding, and select the Z-axis positive direction as the extrusion direction.

The settings of the *Limits* and *Offset* options are shown in Figure 4-86. Click the middle mouse button to complete the creation of the base.

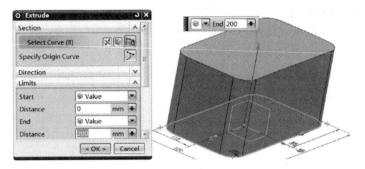

Figure 4-85   Extrude

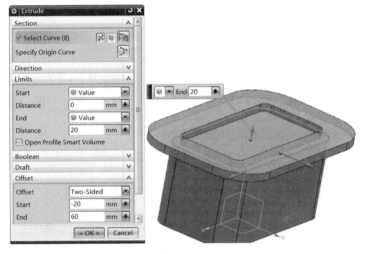

Figure 4-86   Extrude option settings

3. Create a shell

The shell is used to hollow out a solid body or to create a shell around it by specifying wall thicknesses.

Select the menu command *Insert > Offset/Scale > Shell* or click the *Shell* button on the feature toolbar to show the *Shell* dialog box. Set the thickness of the shell to 20, and select the face where the cursor is located in Figure 4-87 as the face to be removed in the graphics area. Click the middle mouse button to create the shell as shown in Figure 4-88.

Figure 4-87   Shell Feature Settings   Figure 4-88   Shell

4. Create a baseplate

(1) Select the menu command *Insert > Sketch*, select the bottom face of the base as shown in Figure 4-88 as the sketch plane, select the X axis positive direction as the horizontal orientation of the sketch, click the middle mouse button to enter Sketch environment.

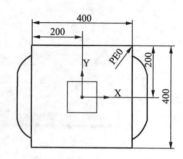

(2) Create the sketch as shown in Figure 4-89 and click the *Finish Sketch* on the toolbar to exit the sketch environment.

Figure 4-89 Sketch of the bottom plate section

(3) Click the *Extrude* on the feature toolbar to bring up the *Extrude* dialog box. In the *Section* option, select the sketch created in the previous step and select the negative direction of the Z axis as the extrusion direction. The setting of the *Limits* option is shown in Figure 4-90. Click the middle mouse button to complete the creation of the case bottom plate.

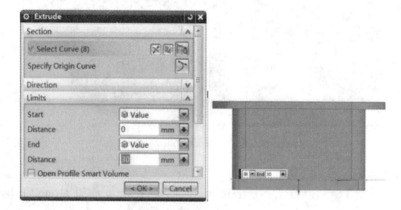

Figure 4-90 Baseplate Extrude Feature

5. Create a support for bearing holes

(1) Select the menu command *Insert > Sketch*, select the plane where the cursor is located as the sketch plane, select one edge of the bottom plate as the horizontal orientation of the sketch, as shown in Figure 4-91. Click the middle mouse button to enter the sketch environment.

(2) Create the sketch as shown in Figure 4-92 and click the *Finish Sketch* on the toolbar to exit the sketch environment.

(3) Click the *Extrude* on the feature toolbar to bring up the Extrude dialog box. In the *Section* option, select the sketch created in the previous step and select the face where the cursor is located as reference for the *Limits* options as shown in Figure 4-93. Click the middle mouse button to complete the extrude feature.

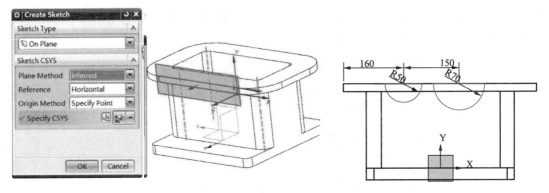

Figure 4-91 Sketch settings   Figure 4-92 Section sketch

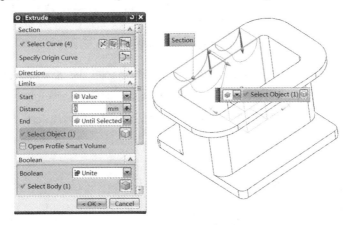

Figure 4-93 Extrude Feature

(4) Select the menu command *Insert > Sketch*, the sketch settings as shown in Figure 4-91. Click the middle mouse button to enter the sketch environment. Create the sketch shown in Figure 4-94 and click the *Finish Sketch* button on the toolbar to exit the sketch environment.

(5) Click the *Extrude* on the feature toolbar to bring up the *Extrude* dialog box. In the *Section* option, select the sketch created in the previous step. The settings for the other options are shown in Figure 4-95. Click the middle mouse button to complete the creation of the support.

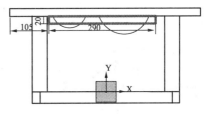

Figure 4-94 Sketch of the extended section

6. Create bearing holes

(1) Select the menu command *Insert > Sketch*, select the plane where the cursor is located as the sketch plane. Select one edge of the bottom plate as the horizontal orientation of the sketch as shown in Figure 4-91. Click the middle mouse button to enter the sketch environment.

(2) In the sketch environment, draw two circles with diameters of 100 and 70,

respectively, centered on the center of the bearing holes support. Exit the sketch environment by clicking the *Finish Sketch* button on the toolbar.

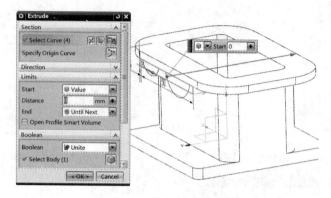

Figure 4-95　Extrude Feature

(3) Click the *Extrude* on the feature toolbar to bring up the *Extrude* dialog box. In the *Section* option, select the sketch created in the previous step. The settings for the other options are shown in Figure 4-96. Click the middle mouse button to complete the creation of the bearing hole.

7. Create a hole

(1) Select the menu command *Insert > Design Feature > Hole* or click the *Hole* on the feature toolbar to show the *Hole* dialog box.

(2) In the *Hole* dialog box, in the *Position* option, specify the point where the center of the hole is located by *Sketch Section*, select the face where the cursor is located as the sketch plane, and select one edge of the bottom plate as the horizontal orientation of the sketch shown as Figure 4-97. Click the middle mouse button to enter the sketch environment.

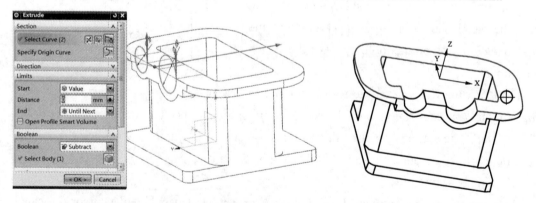

Figure 4-96　Bearing Hole Extrude Option Settings　　　Figure 4-97　Hole Position Sketch Settings

(3) Create the sketch as shown in Figure 4-98 and click the *Finish Sketch* on the toolbar to exit the sketch environment.

(4) In the *Hole* dialog box, select *Type* as *General Hole*; *Form* as *Simple*; the dimensions

of the hole are shown in Figure 4-99. Click the middle mouse button to complete the creation of the hole.

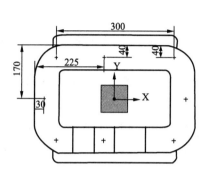

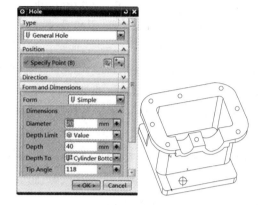

Figure 4-98 Hole position sketch

Figure 4-99 Hole parameter setting

8. Create counter bore

(1) Select the menu command *Insert > Design Feature > Hole* or click the *Hole* on the feature toolbar to show the *Hole* dialog box.

(2) In the *Hole* dialog box, in the *Position* option, specify the point where the center of the hole is located by *Sketch Section*, select the face where the cursor is located as the sketch plane and select one edge of the bottom plate as the horizontal orientation of the sketch as shown in Figure 4-99. Click the middle mouse button to enter the sketch environment.

(3) Create the sketch shown in Figure 4-100 and click the *Finish Sketch* on the toolbar to exit the sketch environment.

(4) In the Hole dialog box, select *Type* as *General Hole*; Form as *Counterbored*; the dimensions of the hole are shown in Figure 4-101. Click the middle mouse button to complete the creation of the hole.

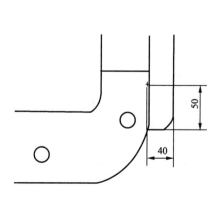

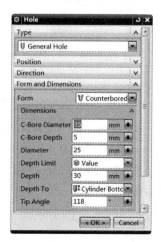

Figure 4-100 Hole position sketch

Figure 4-101 Hole parameter

9. Create threads

(1) Select the menu command *Insert > Design Feature > Thread* or click the *Thread* on the feature toolbar to show the *Thread* dialog box. The software provides two ways to create threads, symbolic thread and detailed thread.

① Symbolic thread.

This method is used to create symbolic thread. Symbolic threads are indicated by dashed circle on the face or faces that will be threaded, which can be used to represent threads and dimension threads in drawings. This type of thread only produces symbols and does not create a threaded body.

② Detailed thread.

This method is used to create detailed threads. This type of thread is more realistic but due to the complexity of this thread geometry it occupies more equipment resources, making it slower to create and update.

(2) In the *Thread* dialog box, select *Thread Type* as *Detailed*; *Rotation* as *Right Hand*; select the created counter bore; the thread parameters are shown in Figure 4-102. If the direction of the thread does not point to the solid at this time, you can click the *Select Start* button, and select the planar face shown in Figure 4-102 as a new starting location in the pop-up dialog box, then click the *Reverse Thread Axis* to specify the thread direction. Click the middle mouse button to complete the creation of the thread as shown in Figure 4-103.

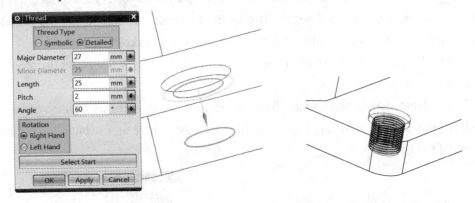

Figure 4-102 Hole feature   Figure 4-103 Detailed thread feature

10. Create a pattern feature

(1) Select the menu command *Insert > Associative Copy > Pattern Features* or click the *Pattern Features* on the feature toolbar to show the *Pattern Feature* dialog box. Select the *Counterbored Hold* and *Threads* features from model tree as the *Feature to Pattern*. In the Pattern Definition option, select *Layout* as *Linear* and set the parameters as shown in Figure 4-104.

(2) Click the middle mouse button to complete the creation of the mounting holes, as shown in Figure 4-105.

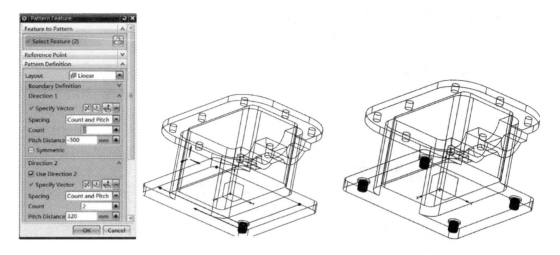

Figure 4-104　Pattern Feature　　　　Figure 4-105　Part after the mounting holes created

11. Create a dart

(1) Select the menu command *Insert > Design Feature > Dart* or click the *Dart* on the feature toolbar to show the *Dart* dialog box.

(2) In the *Dart* dialog box, select the outer cylindrical surface of the bearing hole support as the first set faces; click the middle mouse button to select the outer surface of the base as the second group set faces; set the shape parameters, as shown in the figure 4-106. Click the middle mouse button to complete the creation of the dart.

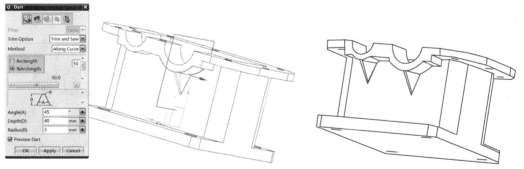

Figure 4-106　Dart Feature　　　　Figure 4-107　Model after the completion of dart

(3) According to the above steps, the dart at the other bearing hole support is completed and its shape parameter is the same as above. The completed model as shown in Figure 4-107.

12. Create mirror features

(1) Select the menu *Insert > Associative Copy > Mirror Feature* or click the *Mirror Feature* on the toolbar to bring up the *Mirror Feature* dialog box.

(2) In the *Mirror Feature* dialog box, select the two extrude features of the bearing hole support, the bearing hole and the two darts as *Features to Mirror*. Select the XY coordinate

plane as the *Mirror Plane* as shown in Figure 4-108. Click the middle mouse button to complete the mirroring of the selected feature.

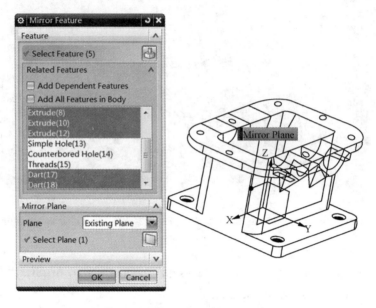

Figure 4-108 Mirror feature settings

Finally save the file and close the current window.

4.8 Exercise - Cylinder Gear Shaft

This exercise will create a cylinder gear shaft part as shown in Figure 4-109. The main structure of the part consists of gear, stepped shafts, thread, chamfers, slot and other features.

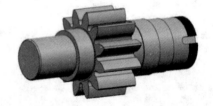

The steps are as follows:

Figure 4-109 Cylindrical gear shaft part

1. Create a multi-diameter shaft

(1) Open the exercise *ch4\example\4-2.prt*, as shown in Figure 4-110.

(2) Create a revolve feature. Select menu command *Insert > Design Feature > Revolve* or click the *Revolve* on the toolbar to show the *Revolve* dialog box.

(3) In the *Revolve* dialog box, select the sketch shown in Figure 4-110 as the section. Select the X-axis as the axis vector, set the starting angle to 0 degree and the ending angle to 360 degree. Click the middle mouse button to complete the creation of the multi-diameter shaft as shown in Figure 4-111.

Chapter 4 | Part Modeling | 109

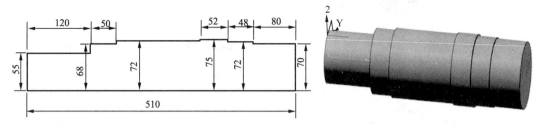

Figure 4-110 4-2.prt   Figure 4-111 multi-diameter shaft

2. Create a key slot

(1) Create a datum plane as the placement face for the slot. Select the menu command *Insert > Datum/Point > Datum Plane* or click the *Datum Plane* on the feature toolbar to show the Datum Plane dialog box. The XZ plane of datum coordinate system and the quadrant point of arc of the end face are sequentially selected as references and the datum plane is established as shown in Figure 4-112.

(2) Select the menu command *Insert > Design Feature > Slot* or click the *Slot* on the feature toolbar to show the *Slot* dialog box, as shown in Figure 4-113. Select the *Rectangular*, click the middle mouse button to show the *Rectangular Slot* dialog box. Select the datum plane made in step 1 as the placement face in the new dialog box, click the middle mouse button to accept the default options.

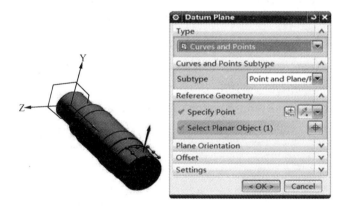

Figure 4-112 Creating a datum plane

(3) Show the *Horizontal Reference* dialog box and select the *Datum Axis* option as the reference object for the horizontal reference. The *Select Object* dialog box is displayed, and the datum axis X axis is selected as the horizontal reference in the graphics area to specify the direction of the slot length.

(4) In the *Rectangular Slot* dialog box, set the slot parameters as shown in Figure 4-114.

(5) Set the positioning dimension along the horizontal direction to 0, and the positioning dimension along the vertical direction to 0.

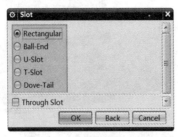

Figure 4-113　*Slot* dialog box

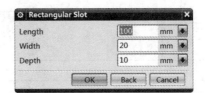

Figure 4-114　*Rectangular Slot* parameters

3. Create a thread

(1) Select the menu command *Insert > Design Feature > Thread* or click the *Thread* on the feature toolbar to show the *Thread* dialog box.

(2) In the *Thread* dialog box, select *Thread Type* as *Detailed*, *Rotation* as *Right Hand*; select the cylindrical face where the slot is located as the reference surface; the thread parameters are shown in Figure 4-115. Click the middle mouse button to complete the creation of the thread.

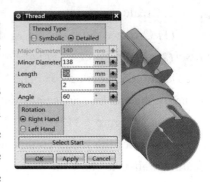

Figure 4-115　Threads parameters

4. Create a cylinder gear

(1) Select the menu command *GC Toolkits > Gear Modeling > Cylinder Gear* or click the *Cylinder Gear* button on the feature toolbar to show the *Involute Cylinder Gear Modeling* dialog box. In the dialog box, select *Create Gear*, click the middle mouse button to bring up the *Involute Cylinder Gear Type* dialog box, then set the parameter as shown in Figure 4-116.

(2) Select the *Addendum Modified Gear* tab and enter the parameters as shown in Figure 4-117. Click the middle mouse button to bring up the *Vector* dialog box.

Figure 4-116　*Involute Cylinder Gear Type* dialog box

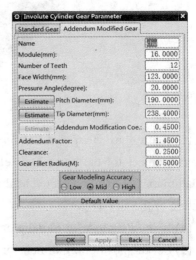

Figure 4-117　*Involute Cylinder Gear Parameter* dialog box

(3) The *Vector* dialog box is used to specify the axis direction of the gear. In the graphics area, select the X axis positive direction as the vector direction of the gear axis as shown in Figure 4-118. Click the middle mouse button to show the *Point* dialog box.

(4) The *Point* dialog box is used to specify the vertex position of the gear. Select the coordinate origin as the vertex position in the graphics area, as shown in Figure 4-119. Click the middle mouse button and the software will automatically complete the creation of the given parameter gear.

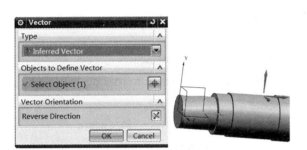
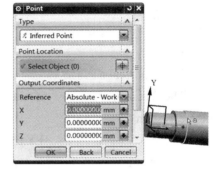

Figure 4-118   *Vector* dialog box   Figure 4-119   *Point* dialog box

4.9 Questions and Exercises

1. Fill in the blanks

(1) Common modeling methods include_____, _____, with_____.

(2) In 3D CAD software, a part model is usually composed of a number of individual geometric elements with certain parameters, just as an assembly consists of many separate parts. These separate geometric elements are called_____.

(3) The 3D model of the part is composed of a series of features constructed in order. If the created features have a datum relationship between them, an association is established between them, that is, it is established_____.

(4) Using features to create parts can be used_____, _____, with_____ Method.

(5) In 3D CAD software, it can usually be divided according to the role of features in the modeling process_____, _____, with_____ feature.

(6) Boolean operations include_____, _____, with_____.

(7) Datum features include_____, _____, _____with_____feature.

(8) In NX, there are two ways to create threads, namely_____with_____.

(9) Linear array method for_____ the form is to copy the selected solid features, and the array pattern is such that the features behind the array are arranged in a rectangle (row number x column number).

(10) The placement surface of the groove feature should be _____.

(11) When you create a slot of the specified length, you need to specify it in addition to the placement plane of the slot_____used to specify the direction of the slot length.

2. Choice

(1) _____is a feature that is generated by revolving a section about a specified axis by a certain angle.

  A. Extrude   B. Revolve
  C. Sweep along the guide line   D. Pipe

(2) For most engineering features, the placement surface is usually_____.

  A. Plane   B. Surface
  C. Cylindrical surface   D. Conical surface

(3) The horizontal datum of the engineering feature is used to define the feature coordinate system-_____.

  A. X axis   B. Y axis   C. Z axis   D. Origin

(4) For rectangular engineering features, _____there are different positioning methods.

  A. 6   B. 7   C. 8   D. 9

(5) In NX, the *Slot* dialog box provides_____types of slot. The slot created can be either a through slot or a slot of a specified length.

  A. 3   B. 4   C. 5   D. 6

(6) When creating a chamfer feature, you need to specify it in addition to its shape parameter_____.

  A. Datum point   B. Datum edge
  C. Datum plane   D. Datum coordinate system

(7) The_____feature is used to remove the material inside the solid and form a thin wall at a specified thickness.

  A. cavity   B. boss   C. shell   D. draft

(8) When you create a face fillet feature, you need to specify_____as a datum in addition to the fillet radius and build method.

  A. Edge   B. Face
  C. Point   D. Coordinate system

3. Short answer questions

(1) What is the difference between the operation of the revolve feature and the operation of the extrude feature?

(2) What is the difference between the operation results of the guide line opening, the guide line closing, the open section, and the section closing along the guide line sweep?

(3) Why do you want to specify the placement surface of the engineering features?

(4) Briefly describe the process of modeling spur gears in the GC toolbox.

(5) What are the two forms of thread features, and what are the differences?

(6) Briefly describe the process of creating a mirrored feature.

(7) Briefly describe the method and process of creating a gear shaft.

(8) Briefly describe the general process of creating triangular ribs.

4. Exercises

(1) Apply the relevant section to create the model shown in Figure 4-120.

Figure 4-120   Model Example

① Open ch4\exercise\4-1.prt, as shown in Figure 4-121, create the first part by extrudeing the feature;

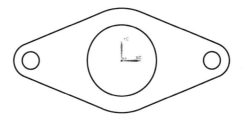

Figure 4-121   4-1.prt section

② Open ch4\exercise\4-2.prt, as shown in Figure 4-122, create a second part with the revolve feature.

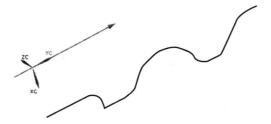

Figure 4-122   4-2.prt section

(2) Open ch4\exercise\4-2.prt, as shown in Figure 4-122, create a model using the revolve feature.

① Establish the revolve feature as shown in Figure 4-123. Take the sketch shown in Figure 4-111 as the section and the datum axis shown in Figure 4-122 as the rotary axis.

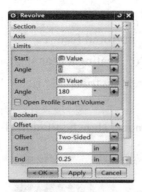

Figure 4-123　Revolving features and settings

② Establish the revolve feature as shown in Figure 4-124. Set the Select Filter Rule option to Face Edge, select the edge of the solid face shown in Figure 4-125 as the profile, and use the short edge of the solid shown in Figure 4-125 as the rotary axis; For 90 degrees, Boolean operations are unite.

Figure 4-124　Revolving features

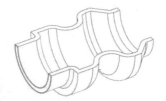
Figure 4-125　Rotary feature setting

(3) Open ch4\exercise\4-3.prt and apply the extrude feature to create the model.

① Establish the extrude feature as shown in Figure 4-126. The section is shown in Figure 4-127. The extrudeing direction is +ZC, the extrudeing limits is 0.5, and the Boolean operation is none.

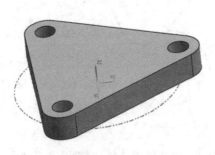
Figure 4-126　Extrude feature

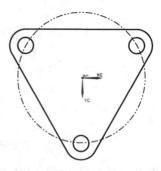
Figure 4-127　Extruded feature section

② Select the upper surface as shown in Figure 4-126, and use it as the sketching plane to draw the sketch shown in Figure 4-128.

③ Create a extrude feature using the sketch shown in Figure 4-128 as a section, as shown in Figure 4-129. The extrudeing direction is +ZC, the extrudeing limits is 2.5, and the Boolean operation is unite.

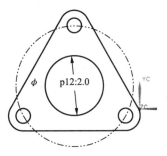

Figure 4-128   Extruded feature sketch   Figure 4-129   Extrude feature

④ Select the upper surface as shown in Figure 4-129, and use it as the sketching plane to draw the sketch shown in Figure 4-130.

⑤ Create a extrude feature using the sketch shown in Figure 4-130 as a section, as shown in Figure 4-131. The extruding direction is +ZC, the extrudeing limits is 0.5, and the Boolean operation is Unite.

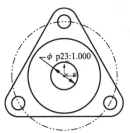

Figure 4-130   Extrude feature sketch (1)   Figure 4-131   Extrude feature (1)

⑥ Select the upper surface as shown in Figure 4-131, and use it as the sketching plane to draw the sketch shown in Figure 4-132.

⑦ Create a extrude feature using the sketch shown in Figure 4-132 as a section, as shown in Figure 4-133. The extruding direction is -ZC, the extruding limits is *Through All*, and the *Boolean* operation is *Subtract*.

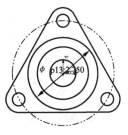

Figure 4-132   Extrude feature sketch (2)   Figure 4-133   Extrude feature (2)

(4) Create a planetary gear reducer housing, as shown in Figure 4-134. The part is mainly used to support the output shaft of the planetary gear reducer and to position the shaft and its

bearing. It consists mainly of hollow cavities in the form of steps in the middle. These stepped structures are used to provide mounting and positioning of the bearings. The large end face of the box is used to connect with the box seat, and there are 4 uniformly threaded holes on the end face.

When you create the part model, you can use the rotary tool to create the base of the box, and then create the threaded holes or create the threaded holes through the array.

Figure 4-134   planetary gear reducer housing

(5) Create a support frame model, as shown in Figure 4-135. This part is primarily used for vertical positioning between shaft parts and other parts. Its main structure consists of countersunk holes on the bottom plate, support plates, cylinders and triangular ribs.

When creating the part, first create the main part of the base plate and the support plate with symmetric extruding, and use the extrude feature to create the groove feature at the bottom. Then use the Hole tool to make the countersunk hole on the bottom plate; then draw the cylinder, and then make the hole feature on the cylinder; finally, use the triangular rib tool to complete the creation of the triangle rib feature, and complete the establishment of the whole model.

(6) Create the oil pump housing parts, as shown in Figure 4-136. The housing is primarily used to protect and secure the pump body system. Its main structure consists of a casing, a base, a bearing hole, and a reinforcing rib.

When creating the part, first create the bottom plate part with the extruding tool, then use the extruding tool to create the shell body, and use the extruding tool to create the cavity; then create the bearing hole base and bearing hole on the end face of the shell, and use the extruding The tool creates the rib body; then it uses the extrude feature to calculate the groove structure of the bottom plate; finally, the part model is perfected with tools such as rounds and threads.

Figure 4-135   Support frame

Figure 4-136   Oil pump housing

Chapter 5

Assembly Modeling

Use the NX Assemblies application to model assemblies of piece part files and subassembly files. An assembly is a part file which contains component objects and a subassembly is an assembly used as a component in a higher level assembly. In this chapter, you will learn how to assemble multiple components or parts into a complete assembly using NX. This chapter mainly introduces NX assembly terms, reference set, assembly constraints, assembly navigator, assembly components edit and so on. The main points of this chapter are:

(1) Understand the basic ideas of assembly design;

(2) Master the assembly constraint command and method of component editing operation;

(3) Be familiar with method of exploded views operation.

5.1 Assembly Overview

5.1.1 Basic process of NX assembly modeling

The assembly file, which is used to record the spatial positional relationship of the component objects, is generated by the NX assembly module.

The assembly in the real world is to connect parts together according to a certain order and some technical requirements to become a complete machine (or product), which must reliably implement the function of the machine (or product) design. The assembly work of the machine generally includes assembly, adjustment, inspection and commissioning, etc. It is not only the final necessary stage to manufacture the machine, but also the total test of the machine design idea, the processing quality of the parts and the quality of the machine assembly. The assembly module of NX comes from the real world, and the complex reality assembly design process is highly reproduced by expressing the assembly relationship between the parts.

The detailed design of each part is carried out under the constraints of the assembly model, and the roughly defined assembly model is refined or modified until the product design meets the requirements.

The NX assembly module provides two basic assembly methods. One is the bottom-up assembly design method, which assembles the parts designed by the modeling module according to a certain order, that is you create parts and then later assemble them together; the other is the top-down assembly design method, which carries out part design using other components as reference and is used in the conceptual design phase of the product, that is you can create geometry at the assembly level, and move or copy the geometry to one or more components. Usually, when assembling the product with NX, these two methods are combined, that is, a hybrid assembly method. For example, create several major component models firstly, then assemble them together, next design other components in the assembly.

5.1.2 Terms in NX assembly

1. Assembly component

An assembly component consists of parts and subassemblies. In NX, it is allowed to add components to any part file to form an assembly, so any part file can be used as an assembly component. It should be noted that when an assembly is saved, the geometric data of each component is not copied in the assembly component file, while is stored in the corresponding component (ie, the part file).

2. Subassembly

A subassembly is an assembly that is used as a component in a higher-level assembly, and the subassembly also has its own components. Subassemblies are a relative concept, and any assembly component can be used as a subassembly in a higher level assembly.

3. Component objects

A component object is a non-geometric pointer to the file that contains the component geometry. After you define a component, the part file in which you define it has a new component object. The component object allows the component to be displayed in the assembly without duplicating any geometry. The information recorded by a component object includes part name, layer, color, line type, line width, reference set, and pairing conditions.

4. Components

A component is a part file that is referred to by a component object in an assembly. A component could be a single component (i.e. a part) or a subassembly. Components are referenced by assembly parts rather than copied into assembly parts.

5. Constraints

A constraint condition, also known as a mating condition, is a specific location condition that needs to be specified when positioning a corresponding component in an assembly. It is usually specified that the constraint relationship between the two components in the assembling process should be fully constrained.

During the assembling process, the user can fix the position of the component by combining different constraints. The relationship between the two components is related, and if the location of the fixed component is moved, the associated subcomponents will also move with it.

6. Master models

A master model is a part file that is commonly referenced by the UG module. Several master model assemblies can reference the same geometry simultaneously, with each assembly file owned by a different individual. You can begin work in the downstream applications before the master model is complete. Because application data is associative to the geometry in the master model, the parts update when the master model is changed.

7. Displayed part

A displayed part is a part that is displayed in the current graphics window.

8. Work part

The work part is the part in which you create and edit geometry, and to which you add components. When the displayed part is an assembly, you can change the work part to any of its components. In a NX assembly, there can only be one work part.

5.1.3 Introduction to the assembly interface

When assembling with NX, you must firstly enter the assembly environment. After starting the NX software, the user can enter the assembly module by creating a new assembly file, or by opening an existing assembly file. The assembly interface is shown in Figure 5-1.

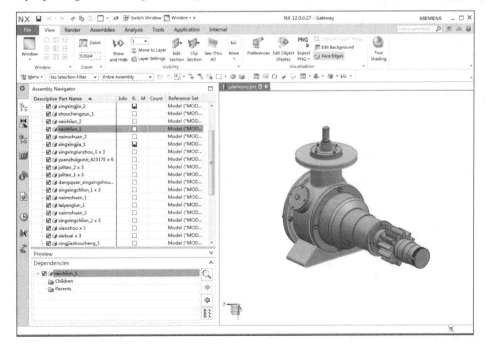

Figure 5-1   assembly interface

The NX assembly interface mainly includes the assembly navigator, assembly toolbars, graphic workspaces, and more. The assembly toolbar is shown in Figure 5-2. The tools in the toolbar can be used to add and create new in the assembly. You can also use the corresponding commands in the *Assembly* menu to achieve the same function. The functions commonly used in the toolbar and how to use them are explained in detail in the following sections, which are not described here.

It provides a set of commands for manipulating components in an assembly environment to add, create new, and pattern component, mirror assembly and more for components in an assembly.

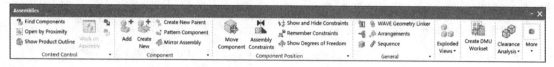

Figure 5-2   assembly toolbar

5.1.4   Assembly Navigator

The assembly navigator, also called the assembly navigation tool, located in the resource bar, which provides a quick and easy way to select components and manipulate components.You can view the parent-child relationship of the assembly parts,modify the name of the assembly, create new parent and create new component with the assembly navigator, besides, modifying the assembly relationship (constraint relationship) of the selected component. In the assembly navigator, the assembly structure is displayed in a tree-like diagram, the so-called "assembly tree", where each component acts as a node of the tree.

1. Open the assembly navigator

Click the *Assembly Navigator* icon on the resource bar, or slide the cursor over the icon to open the assembly navigator, as shown in Figure 5-3.

Figure 5-3   assembly navigator

In the assembly navigator, in order to identify each node, the subassembly and components in the assembly are represented by different icons. At the same time, the icons that indicate the different states of the assembly or component are also different.

(1) Assembly or subassembly: when the icon shows yellow 🗂, it means that the assembly or subassembly is in the work part; if it is gray, but there is a black solid border, it means that the assembly or subassembly is a non-work part; if it is all gray, it indicates that the assembly or subassembly is closed; if it is blue 🗂,it indicates that the assembly has been suppressed.

(2) Component: when the icon shows yellow 🗂, it means the component is inside the work part; if it is gray, but there is a black solid border, it means the component is a non-work part; if it is all gray, it means the component is closed; if it is blue 🗂 , indicating that the component has been suppressed.

(3) Check box: indicates the display status of the assembly or component. If the check box is selected, it will be red ✓, indicating that the current part or assembly is in the display state; if it is gray ✓, it means the current part or assembly is hidden; if it is not selected (single box) ☐, it means the current part or assembly is closed; A dashed box indicates ⊡,that the part has been suppressed. The display effect of each component and its check box in different states, as shown in Figure 5-4.

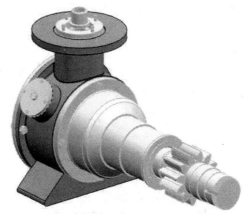

Figure 5-4　different states of the component

(4) Expand ⊞ and compression ⊟ of the assembly tree node: Click the plus sign to expand the assembly or subassembly, display all components of the assembly or subassembly, and once clicked, add the minus sign. Clicking the minus sign indicates that the assembly or subassembly is not displayed, and the subordinate components are not displayed, that is, an assembly or subassembly is compressed into a node, and the minus sign is added to the plus sign.

(5) Preview and Dependencies. Preview: preview the selected components. Dependencies: displays the parents and children of the selected assembly or component node, as shown in Figure 5-5.

Figure 5-5   Preview and Dependencise

2. The right-click menu in the Assembly Navigator

1) right-click menu for node

Select one of the nodes in the assembly navigator and right click to pop up the shortcut menu, as shown in Figure 5-6. The options in this menu vary depending on the state of the component. The commonly used menu functions are as follows:

(1) Make work part: convert the part into a work part, and other parts will be displayed in gray as shown in Figure 5-6, the component daxiang_ti is a work component.

(2) Open in window: convert the component into a displayed part and show it in a new window, as shown in Figure 5-7;

Figure 5-6   Node right-click menu

(3) Replace Reference Set: replace the reference set of the currently selected component. You can replace the selected component with a custom reference set or a system default reference set;

(4) Open: open the component in the assembly tree. If an assembly is already open and its subordinate components are closed, you can open these closed components in this way.

(5) Close: closing the selected component or the entire assembly could improve the speed of operation.

(6) Show Only and Hide: this command is useful when adding assembly constraints. It is shown as a graphic area showing the selected component or assembly visible, hidden as a graphic area showing the selected component or assembly is not visible; after executing the hide command on the selected component. The effect of hide operation is shown in Figure 5-8.

Figure 5-7 Show parts   Figure 5-8 Hide parts

(7) Properties: listing information about the selected component. These informations includes the component name, the associated assembly name, color, reference set, constraint name, and properties.

2) right-click menu for blank area

In any blank area of the assembly navigator, right-click to pop up the shortcut menu, as shown in Figure 5-9.

Select the specified command in the shortcut menu to perform the corresponding operation. For example, when you select *Column >Position* and *Column > Reference Set*, only the position and reference set of the component are displayed in the column of the assembly navigator, as shown in Figure 5-10.

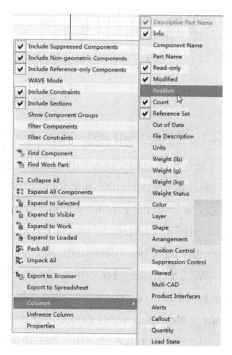

Figure 5-9 Blank area right-click menu

5.1.5 Reference set

In a virtual assembly, it is generally not desirable to refer all the information of each component to an assembly. Usually, only the solid graphics of the component are required, and many components also include other reference information of planes, datum axes, sketches, etc, which takes up a lot of memory and can cause unnecessary trouble to the assembly. Therefore, NX allows the user to select a part of the geometric object as a representative of the component to participate in the assembly, which is the reference set.

Each component created by the user contains a default reference set. There are three default reference sets: model, empty, and the entire part. In addition, the user can modify and create the reference set. Select the reference set command in the drop-down menu, and pop up the *Reference Set* dialog box shown in Figure 5-11, which provides the function of creating, deleting, and editing the reference set.

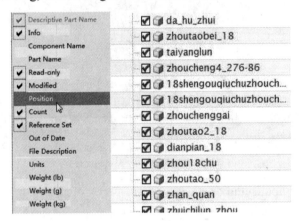

Figure 5-10 Setting of column   Figure 5-11 *Reference Set* dialog box

5.2 Assembly Method

5.2.1 Bottom-up assembly modeling

In a bottom-up assembly modeling, the entire assembly model of the product is obtained by assembling designed individual components or subassemblies which are added to assembly model to form the assembly. It is a common assembly method used in assembly modeling, and is suitable for situations where the assembly relationship is clear and the components in the assembly have been designed.

When creating bottom-up assembly, it is common to load the designed components into the assembly environment, then place them by setting their constraints, and effectively manage the components data by setting their reference set.

To add a component that has already been created, you can select the *Add* button in the assembly toolbar or select the menu *Assemblies* > *Components* >*Add Component* command, and the *Add Component* dialog box will pop up, as shown in Figure 5-12. In this dialog box, the user can open or specify the part to be added to the assembly, and set its location, constraint, reference set, preview and so on.

1. Part To Place

In this dialog, there are 3 ways to specify existing parts, namely:

(1) Select part: click the *Select Part* button, and then select the corresponding part in the graphics area as the specified part;

(2) Loaded parts: in the *Loaded Parts* list box, the system automatically collects the loaded parts into the list box, and the user can select the corresponding part as the specified part;

(3) Open part: click the *Open* button to pop up the *Part Name* dialog box. The user can specify the part path and then select the corresponding part as the specified part.

2. Location

In NX, the system provides 3 ways to set the location, namely:

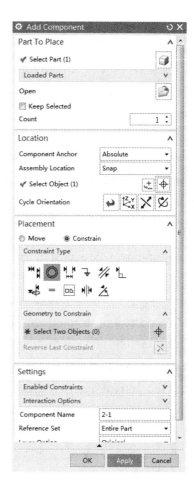

Figure 5-12 *Add Component* dialog box

(1) Absolute origin: absolute origin location, which will make the working coordinate system of the component to be placed coincident with the working coordinate system in the assembly environment.

Usually, only one part can be placed in this way in the assembly model, the first added component is placed in this location method.

(2) Select the origin: the system determines the location of the component by specifying the origin. The origin of the coordinate system of the component to be placed coincides with the selected origin. After selecting this option, click the middle mouse button to pop up the *Point* dialog box, the user specifies the location of the point in the dialog box to complete the placement of the component.

(3) Constraint: The placement of the component is done by selecting the component reference and setting the corresponding constraint type. The specific constraint types will be described in details later.

3. Reference set

The reference set option is used to set the reference set for the specified part when it is added to the assembly.

4. Preview

After the preview option is selected, the *Preview* dialog box will pop up, and the corresponding component will be shown according to the reference set option setting by the user. After the user selects the placement type and the like, it can also assist the user to complete the selection of the positioning reference. As shown in Figure 5-13.

Figure 5-13　*Component Preview* dialog box

5.2.2　Top-down assembly modeling

In top-down assembly modeling, the assembly structure is usually represented by some datum objects, the assembly relationships between the components are controlled by these datum objects, the shape or location determined by previously assembled components.

When creating a top-down assembly, it is common to create a new component in the current assembly, and use this component as work part, then proceed with the structural design of the component. When designing a component structure, it is usually necessary to associate it with the current assembled component, that is, to link the object to the current assembly environment using the WAVE geometry linker.

1. Create New Component

To create a new component in an existing assembly, you can select the *Create New* button in the assembly toolbar or select the menu *Assemblies > Components > Create New Component* command to bring up the *New Component File* dialog box. Just like the *New File* dialog box, select the component type in the dialog box, and enter the component file name and the save path of the component file, then click the middle mouse button to pop up the *Create New Component* dialog box, as shown in Figure 5-14. In the dialog box, click the middle mouse button to complete the creation of the new component.

Figure 5-14　*Create New Component* dialog box

After the new component is generated, since it does not contain any geometric objects, the user needs to establish its geometry. Therefore, the user should set the new component as the work part, then complete the creation of its model. Usually, there are two kinds of methods to build new component geometric objects:

1) Create non-associative geometric objects

The geometric object is modeled directly in the new component, then it will be placed to the specified location. Since there is no reference to the geometry of the existing component during the modeling of the new component, there is no associative links between the dimensions and other components.

2) Create associative geometry objects

Different from the previous method of the new component modeling, this modeling method is usually required that the new component has a geometric relationship with other components in the assembly, that is, a geometric link relationship will be established between the components. In NX, WAVE geometry linker technology is used to achieve this relationship.

2. WAVE Geometry Linker

The WAVE technology used by NX is a correlation parametric design technique based on assembly modeling, which allows to establish the geometry link between different components, the so-called "inter-part correlation" relationship, thus achieving the associative copy of objects between components.

The main way to establish a link relationship between components is to use the WAVE Geometry Linker. Set the new component as the work part, then select the *WAVE Geometry Linker* button in the assembly toolbar to pop up the *WAVE Geometry Linker* dialog box, as shown in Figure 5-15.

Figure 5-15 *WAVE Geometry Linker* dialog box and Type option

This dialog is used to link objects such as curves, faces, and bodies of other components to the current work component. The specific meanings of the geometry types are as follows:

(1) Composite Curve : this option is used to select specified curves or edges from other components to create a linked curves and link the selected curves to the specified work part.

(2) Point : this option is used to pick a point from other components to establish a link point.

(3) Datum : this option is used to select the corresponding datum plane or datum axis from other components to establish a link datum.

(4) Sketch 📷 : this option is used to select the appropriate sketch from other components to create a linked sketch.

(5) Face 📷 : this option is used to select one or more specified solid surfaces from other components to create a link face. Figure 5-16 shows the extrude features created by the user using the edge of the link surface as sketch section. After the link surface is modified, the extrude feature with this link surface as a reference is also automatically updated, as shown in Figure 5-17.

(6) Region of Faces 📷 : this option is used to select the specified seed face and boundary face from other components to create a link area surrounded by the specified boundary.

(7) Body 📷 : this option is used to select the specified entity from other components to create a link body.

(8) Mirror Body 📷 : this option is used to select the specified entity from other components, then specify the mirror plane to establish the link of the specified entity.

(9) Routing Object 📷 : this option is used to select the corresponding object from other components to create a routing object.

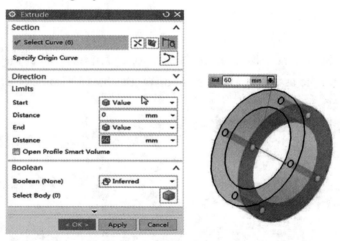

Figure 5-16   Link surface application

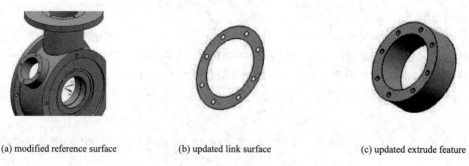

(a) modified reference surface   (b) updated link surface   (c) updated extrude feature

Figure 5-17   Link surface relevance

5.3 Assembly Constraints

Any component in the space includes 6 degrees of freedom, three translations and three rotations. Assembly modeling is the process of specifying the relative location and orientation of parts, which is applying constraints to constrain geometric entities between mating parts.

In the process of applying assembly constraints, it usually involves two components: a mating component and a base component, where the mating component refers to a component that needs to be positioned, and the base component refers to a component whose location has been constrained.

The user can set the placement method of the component in the bottom-up assembly by *Constrain* or select the *Assembly Constraints* button on the assembly toolbar or select the menu *Assemblies>Component Position>Assembly Constraints* command, pop-up *Assembly Constraints* dialog box, as shown in Figure 5-18.

5.3.1 Constraint status

Constraint play a linking role in component assembly. According to the number of constraints applied to the assembled part, it can be divided into the following 4 constraint status:

(1) Fully Constrained ●: 6 degrees of freedom (DOFs) of the component are all comstrained, indicating that the component is fully constrained, and can not move at all; the component that reaches the fully constrained state can be regarded as a rigid body during kinematic analysis.

(2) Underconstrained ◐: the degree of freedom of restriction is less than 6; indicating that the component is partially constrained, and also allowed to move with respect to one another.

(3) Unconstrained ○: indicating that the component is unconstrained and can be moved freely.

(4) Constraints conflict ✖: indicating there are conflicts with existing defined constraint. For example, the same degree of freedom is over-constrained by multiple restrictions.

Usually when imposing constraints, we should avoid inconsistent constraints and unconstrained.

After defining the constraints, you can select the *Show Degrees of Freedom* button on the assembly toolbar or select the menu *Assemblies > Component Position >Show Degrees of Freedom* command to display the constraint symbol in the graphics window, as shown in Figure 5-19, where translationanl DOFs indicate that the part is allowed to translate along certain directions, and the rotational DOFs indicate that the part is allowed to rotate about the axis. If the component is completely constrained, no constraint symbols are visible in the graphics window.

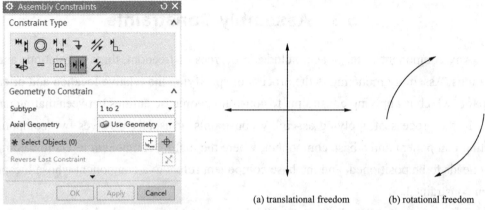

Figure 5-18　*Assembly Constraints* dialog box　　　　Figure 5-19　Constraint symbol

5.3.2　Fix

The Fix constraint ⊥ is used to fix the component at its current location. Usually the first component brought into the assembly is fixed to the default datum features with all DOFs constrained.

5.3.3　Touch Align

Touch Align is the most commonly used constraint. In NX, the touch constraint and the align constraint are combined into one constraint type, that is the Touch Align constraint ⋈. Several placement methods of this constraint are described below:

1. Prefer Touch

The prefer touch orientation to let NX calculate the constraint type.

2. Touch

Touch constraint ⋈ are used to align the normals of two similar objects of constrained components. For example, the normal directions of the two planar objects are opposite or the axis vectors of the holes are antialigned and coincident.

When setting a touch constraint, two planes or faces are specified as constraint references, and the two planes or faces are coplanar and the normal direction is opposite. As shown in Figure 5-20, the highlighted plane in the figure is the specified plane, and the vertical arrow of the plane indicates the normal direction of the plane, the other arrows indicate the degree of freedom of the selected component. As a result, the translational motion of the component along the normal direction of the two planes is constrained, and the DOFs along the other directions remain. For the conical shaped object, the system firstly checks whether the angles are equal. If they are equal, align the axis, as shown in Figure 5-21. For curved objects, the system firstly checks whether the inner and outer diameters of the two faces are equal. If they are equal, align the axis and positions of the two faces. For cylindrical objects, the two cylindrical faces that require mutual fit should be equal in diameter to align the axis.

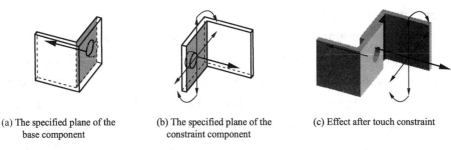

(a) The specified plane of the base component   (b) The specified plane of the constraint component   (c) Effect after touch constraint

Figure 5-20   Plane touch constraint

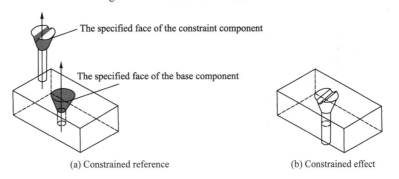

(a) Constrained reference   (b) Constrained effect

Figure 5-21   Conical surface touch constraints

3. Align

The align constraint ▣ is used to make the two similar objects placed in the same direction. When setting the align constraint, two plane objects are specified as constraint references, the two planes are coplanar and have the same normal direction, as shown in Figure 5-22. The highlighted plane in the figure is the specified plane, and the vertical arrow of the plane indicates the normal direction of the plane, the other arrows indicate the degree of freedom of the selected component, the translational motion of the component along the normal direction of the two planes is constrained; when the specified cylinder, cone or the torus is a constrained reference, its axis will be consistent, but their diameters are not required to be the same; when the edges and lines are aligned, the two are collinear.

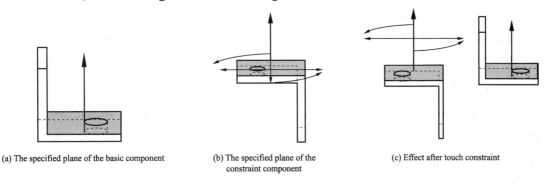

(a) The specified plane of the basic component   (b) The specified plane of the constraint component   (c) Effect after touch constraint

Figure 5-22   Plane align constraint

4. Infer Center/Axis

Infer center/axis constraint is referenced to the specified cylindrical or conical surface, the system will automatically use the center or axis of the face instead of the face itself as a constraint object. As shown in Figure 5-23, the highlighted cylindrical surface in the figure is the specified reference, the black arrow indicates the axis of the cylindrical surface, and the other arrows indicate the degree of freedom of the selected component. The translational motion of the vertical axis of the component is constrained. There are degrees of freedom in other directions.

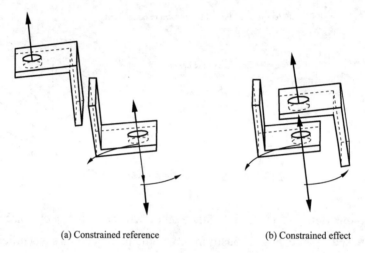

(a) Constrained reference   (b) Constrained effect

Figure 5-23   Cylindrical surface infer center/axis constraint

5.3.4 Distance

The distance constraint is used to specify the minimum distance between two components. The distance can be positive or negative. The sign is used to determine which side of the base component the constraint component will be on.

5.3.5 Concentric

Concentric constraint specifies the circular or elliptical edges of two components as references to coincide the centers of the selected references and make the planes of the edges coplanar.

5.3.6 Center

The center constraint centers one or two objects between a pair of objects, or centers a pair of objects along another object. When making a center constraint setting, you should select the reference object on the mating component and the base component in turn according to the system cues.

This constraint includes the following subtypes:

(1) 1 to 2: position an object in the mating component to the center of symmetry of the two objects in the base component. As shown in Figure 5-24, select the cylindrical surface as the

reference for the constrained component. The two sides of the rectangular groove are used as the reference for the base component, and the center of the cylindrical surface is aligned with the center of the groove.

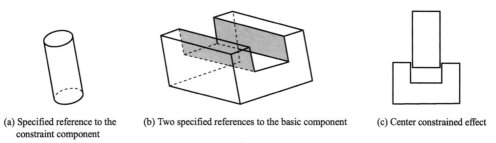

(a) Specified reference to the constraint component   (b) Two specified references to the basic component   (c) Center constrained effect

Figure 5-24   Center constraint

(2) 2 to 1: position the symmetric center of the two objects in the constrained component to the center of an object in the base component.

(3) 2 to 2: make two objects in the constrained component symmetrically arranged with two objects of the base component.

5.3.7   Angle

Angle constraint ∡ is the definition of an angle dimension between two objects that constrains the selected component to the correct orientation. The angle constraint can be generated between two objects with direction vectors, the angle is between the two direction vectors, and the counterclockwise direction is positive. Angle constraints allow you to select different types of objects, for example, you can specify an angle constraint between face and edge.

5.3.8   Parallel and Perpendicular

Parallel ∥ and perpendicular ⊥ constraints are orientation constraints, they only define the geometric relationships of the specified reference objects and do not restrict the location between their components. The parallel constraint makes the direction vectors of the two objects parallel to each other, and the perpendicular constraint makes the direction vectors of the two objects perpendicular to each other.

5.3.9   Fit and Bond

The fit constraint = is used to combine two cylindrical faces of equal radii. This constraint is useful for determining the location of the pin or bolt in holes. If the radii become non-equal in the future, the constraint is invalid.

Bond constraint is used to "weld" the components together so that they can move like a rigid body. Bond constraints can only be applied to components, or a geometry of components and assembly-level, and other objects are not optional.

5.3.10　Align/Lock constraint

Align/Lock is used to align two axes in different objects while prevent rotating around the common axis.

5.4　Component Edit

During the assembly process, it is necessary to perform editing operations such as pattern, mirror and replace on the components sometimes. The functions of the component pattern, mirror, etc. described in this section are similar to those of the modeling module. The objects operated in the modeling module are features, and the objects operated in the assembly module are parts or subassemblies.

5.4.1　Mirror

The component mirror is used to handle assemblies with symmetric component. Component mirror can be used to assemble another instance of a specified part (copy) in a mirrored relationship, or create a mirrored part of a specified part at a certain planar position. Mirrored assembly can reduce cumbersome assembly operations and quickly get new reflected components. By mirroring the components, you can maintain a mirror-symmetric relationship between the source and mirrored parts. The mirrored part allows the user to create a new copy of the part or assembly. The new mirrored the geometry and position of the model can be associative or non-associative to the original model.

For associative copies, if the original component changes, the copied or mirrored component also changes. The mating between the original parts can be saved in copied or mirrored parts. The configuration in the original component appears in the copied or mirrored component.

The mirror feature in the modeling module and the mirror assembly in the assembly module have similarities, and should carry out selection (features selection or parts selection). The difference between them is that the mirror feature refers to the feature, geometry constraints and size constraints, the mirror assembly refers to the assembly constraints between components.

To mirror selected components, the user can choose the *Mirror Assembly* button in the assembly toolbar or select the menu *Assemblies> Components >Mirror Assembly* command to pop up the *Mirror Assemblies Wizard* dialog box, as shown in Figure 5-25. Click the *Next* button, and select the component to mirror in the opened dialog box. It should be noted that only the components of work part can be selected to mirror. As shown in Figure 5-26.

Chapter 5 | Assembly Modeling | 135

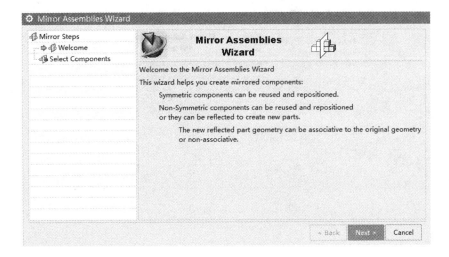

Figure 5-25   General introduction

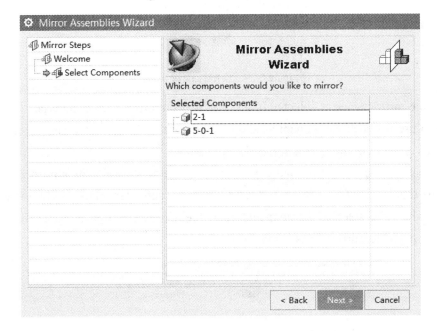

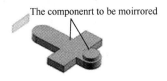

Figure 5-26   Select components

Next click the Next button, select or create the plane on which the assembly will be mirrored, as shown in Figure 5-27. If there is no suitable plane, you can click the *Create Datum Plane* button to opens the *Datum Plane* dialog box where you can create a datum plane or modify an existing datum plane.

Figure 5-27   Select mirror plane

After the mirror plane is selected, click the *Next* button to define the type of mirror that you want to use for each component as shown in Figure 5-28. The user can select the mirror component, and click the *Associative Mirror* button or the *Non Associative Mirror* button to create new part files that contain associative or nonassociative mirrored geometry. At the same time, the *Reuse and Reposition* button is activated, click this button to create a new instance of the selected component; click the *Exclude* button to exclude selected components from the mirrored assembly.

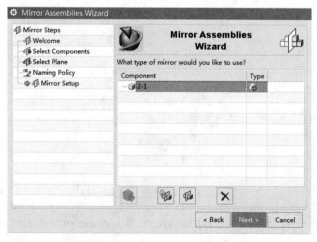

Figure 5-28   Mirroring/Mirroring Settings

Next click the *Next* button to review any of the default actions that you defined on the

Mirror Setup page. As shown in Figure 5-29, in this dialog box, you can select multiple rows to make a change to several components at the same time. If you are not satisfied with the previous results, you can click the *Cycle Reposition Solutions* button ⊙ to cycle through each possible reposition solution for the selected component, or you can select a solution from the list box.

After determining the mirroring result, click the *Next* button to name the newly-created part file as shown in Figure 5-30. After confirming the file name, click the *Finish* button to complete the creation of the mirror component, as shown in Figure 5-31.

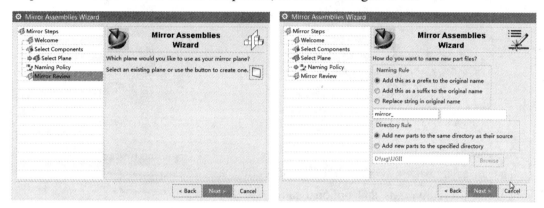

Figure 5-29　Mirroring /viewing　　　　Figure 5-30　Mirroring/naming

Figure 5-31　Component mirror result

5.4.2　Pattern

The pattern component is used to create copies of components and lay them out in a pattern formation.

To pattern component, the user can select the *Pattern Component* button in the assembly toolbar or select the menu *Assemblies>Components>Pattern Component* command to pop up the *Pattern Component* dialog box as shown in Figure 5-32. In the dialog box, you can select one or more components to pattern. After specifying the pattern components, you should set the pattern layout type from *Layout* listbox.

1. Linear

The linear pattern component layout defines a layout using one or two directions. To do this, choose the *Linear* option in Figure 5-32, then to define linear directions by specifying vector as shown in Figure 5-33, then to set the spacing methods including the number of objects and distance between copies of the selected components in the pattern in specified direction.

Figure 5-32　*Pattern Component* dialog box

Figure 5-33　*Vector* setting

2. Circular

The circular pattern component layout define a layout using a rotation axis and optional radial spacing parameters. To perform this operation, select the component for pattern, then select the *Circular* type from *Layout* listbox in the *Pattern Component* dialog box as shown in Figure 5-34.

The circular pattern can be completed by setting the specified vector and the specified point for the rotation axis and the *Angular Direction* parameters. As shown in Figure 5-35, select the surface of hole as reference for rotation aixs. Then set the count of 3 and span angle of 120 degree, click *OK* to complete the circular pattern of selected component, the result is shown in Figure 5-36.

Figure 5-34　*Circular Pattern* dialog box

Figure 5-35　Rotation axis reference

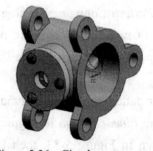

Figure 5-36　Circular pattern result

5.5 Exploded Views

The user can use exploded view to create a view in which selected parts or subassemblies are moved apart visually. An assembly modeling can generate multiple exploded views by adjusting the separation distance between the parts. As a view, once exploded view has been defined and named, it can be referenced in other modules.

Exploded views can be generated at any level of the assembly model. The exploded view is associated with the displayed component and saved in the displayed component. The user can use the explosion in any view and can edit this explosion view.

The user can select the *Exploded Views* button in the assembly toolbar to pop up the Exploded Views toolbar, as shown in Figure 5-37.

Figure 5-37 *Explosion Views* toolbar

5.5.1 Create exploded views

1. New explosion

Use the *New Explosion* button to create a new explosion, in which components can be visually repositioned to produce an exploded view. The *New Explosion* dialog box will pop up, as shown in Figure 5-38, you specify a name for the explosion or use the default name in the *Name* text box, and click the middle mouse button to accept this name. After you create a new explosion, the next step is to edit it. Use the *Auto-explode Components* or *Edit Explosion* command to achieve that.

Figure 5-38 *New Explosion* dialog box

2. Auto-explode components

The Auto-explode components to define the position of one or more selected components in an exploded view. This command offsets each selected component along a normal vector based on the assembly constraints of the component.

Make sure the work view displays the explosion you want to edit. Then user can select the *Auto-explode Components* button in the *Exploded Views* group to pop up the *Class Selection* dialog box. Then select the components you want to offset. Click the middle mouse button to open the *Auto-explode Components* dialog box for specifying the offset value as shown in Figure 5-39.

The *Distance* option is used to set the offset distance for the exploded components. The direction of the automatic explosion is controlled by the input value which can be positive or negative.

Figure 5-39   *Auto-explode Components* dialog box

3. Edit explosion

Explosive vision generated by Auto-explode components generally does not achieve the desired explosion effect, and the explosion views usually need to be adjusted.

The user can use the *Edit Explosion* command to reposition one or more selected components in an explosion. This command is available when the work view displays an explosion.

Make sure the work view displays the explosion you want to edit. Choose *Edit Explosion* in the *Exploded Views* group to pop up the *Edit Explosion* dialog box as shown in Figure 5-40.

Figure 5-40   *Edit Explosion Views* dialog box

Make sure *Select Objects* is selected, then select the component you want to move in the graphic area, then the selected component will be highlighted, as shown in Figure 5-40. Next select *Move Objects*, then you can move or rotate the component to a desired location through dragging the origin handle as shown in Figure 5-41.

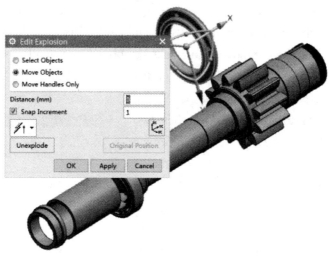

Figure 5-41   *Edit Explosion* setting

5.5.2   Unexplode component and Delete explosion

1. Unexplode component

The user can select the *Unexplode Component* button in the *Exploded Views* group to pop up the *Class Selection* dialog box. Then select one or more componentst you want to restore them to their original, unexploded position. After specifying the components, click the middle mouse button and the component will return to the original location.

2. Delete explosion

The user can select the *Delete Explosion* button in the *Exploded Views* group to pop up the *Explosion Views* dialog box, as shown in Figure 5-42. If more than one exploded view exits, a list of all exploded views appears which lets you select the view to delete. Then select the view and click the middle mouse button to delete the view with the specified name.

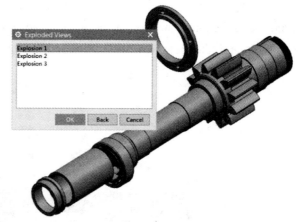

Figure 5-42   *Exploded Views* dialog box

5.6 Assembly Interference Check

When an assembly is completed, interference check is a necessary step. If this check is not carried out, it is very likely that the parts will not be assembled during the practical production.

Assembly interference refers to the phenomenon of volume intersection between parts. Interference in the parts will cause parts collision and parts will not be installed correctly. For the moving mechanism, the components of the assembly are constantly moving, so it should make sure that no interference occurs between parts at each location.

The user can select the menu *Analysis>Simple Interference* command to determine whether two bodiesint ersect. Choose *Menu>Analysis>Simple Interference* to pop up *Simple Interference* dialog box. Select two components you want to check for interference as the first body and the second body, respectively. Select the output form of the interference check result by options of *Highlighted Face Pairs* or *Interference Body*. After the setting is completed, click the middle mouse button to perform the check and output the check result, as shown in Figure 5-43.

Figure 5-43 *Simple Interference* dialog box

5.7 Exercise - Planetary Reducer Output Shaft Assembly

The output shaft of the planetary reducer is a gear shaft, the left end is a spline connected with the spline groove of the planet carrier; the right end is an output end, and the flat key groove and the pulley are connected with the actuator. The middle part is a bearing at a certain distance, and the bearings are positioned by the inner ring of the sleeve; as shown in Figure 5-44.

Figure 5-44 Assembly model of planetary reducer output shaft

The planetary reducer output shaft assembly adopts bottom-up assembly method, and the assembly process is as follows:

(1) New file: open NX software, create a new assembly file, the file name is *outputshaft*, click the middle mouse button to enter the assembly environment.

(2) Load the output shaft part: select the *Add* button in the assembly toolbar or select the menu *Assemblies>Components>Add Component* command, pop up the *Add Component* dialog box. Click *Open* in the dialog box, select *shuchuzhou* file in the folder *ch5\example*, set *Assembly Location* to *Snap*, as shown in Figure 5-45. Click *Select Object* in *Location* group, in the pop-up *Point* dialog box click *OK* to accept default point position; Click The middle mouse button to accept the position of the output shaft part;

(3) Output shaft part placement: click the *OK* button in the dialog box in Figure 5-46 to fix the output shaft at the current location.

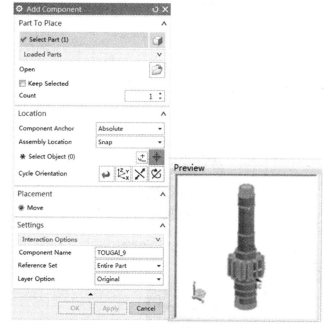
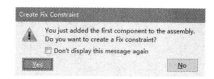

Figure 5-45   Loading the output shaft part   Figure 5-46   Create Fix Constraint Dialog Box

(4) Load the sleeve: choose the *Add* button to pop up the *Add Component* dialog box, click in the dialog box, and select *youzhoutao_9* in the folder *ch5\example*; set *Placement* to *Constrain*; select *Touch Align* as *Constraint Type*, set *Orientation* as *Preferred Touch*, and then sequentially select the end faces on the sleeve and the output shaft as references as shown in Figure 5-47. The system performs the touch constraint operation automatically. Continue to select the *Constraint Type* as *Touch Align*, the *Placement* method is *Infer Center/Axis*, and then select the outer cylindrical surface of the sleeve and the cylindrical surface on the output shaft as references as shown in the figure 5-48; the effect of the shaft assembly is shown in Figure 5-49.

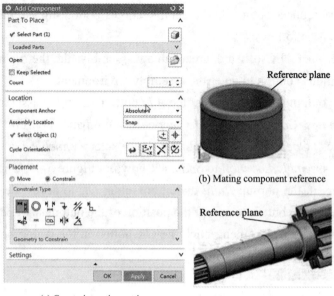

Figure 5-47　Sleeve constraint setting

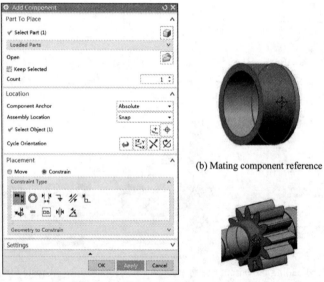

Figure 5-48　Sleeve constraint settings

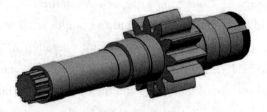

Figure 5-49　Result of the sleeve constraint

(5) Load the cover assembly: select the *Add* button to pop up the *Add Component* dialog box, click in the dialog box, select *outputsub2ass* file in the folder *ch5\example*; Set *Placement* to *Constraints*, set constraint type as *Touch Align* and the *Orientation* type as *Align*, as shown in Figure 5-50. Then select the end face of the cover assembly and the end face of the sleeve as the references, that is, the face where the cursor is located in the figure. The constrained operation will be performed. Continue to select the constraint type as *Touch Align*, the *Orientation* method is *Infer Center/Axis*, select the outer cylindrical surface of the cover assembly and the cylindrical surface of the sleeve, and the effect of the shaft assembly is as shown in Figure 5-51.

(a) Mating component reference   (b) Base component reference

Figure 5-50   Cover constraint setting   Figure 5-51   Cover constraint result

(6) Load the bearing: click the *Add* button, and open the *zhoucheng_11* file under the folder *ch5\example*, and set the *Placement* method to *Constraint*. Select the constraint type as *Touch Align*, the *Orientation* method is *Preferred Touch*, and then select the bearing end face and the cover assembly end face as references as shown in Figure 5-52. Continue to select the constraint type as *Touch Align*, the *Orientation* method is *Infer Center/Axis*, and then sequentially select the outer cylindrical surface of the bearing and the cylindrical surface of the through-cover assembly. The effect of bearing assembly constraint as shown in Figure 5-53.

(a) Mating component reference   (b) Base component reference

Figure 5-52   Bearing Constraint Settings   Figure 5-53   bearing constraint result

(7) Load the bushing: choose the *Add* button, and then open the *youbanbu_zhouchengtao* file under the folder *ch5\example*, set the Placement mode to *Constraint*. Select the constraint type as *Touch Align*, the *Orientation* method is *Preferred Touch*, and then select the sleeve end face and the end face of bearing inner ring as the

references in turn, as shown in Figure 5-54. Continue to select the constraint type as *Touch Align*, the *Orientation method* is *Infer Center/Axis*, then select the cylindrical surface of the sleeve and the outer cylindrical surface of the bearing as the references in turn, and the effect of bearing assembly is shown in Figure 5-55.

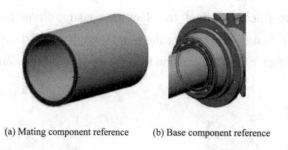

(a) Mating component reference　　(b) Base component reference

Figure 5-54　Bearing Constraint Settings　　Figure 5-55　Bushing constraint result

(8) Move the bearing: select the *Move Component* button, in the *Move Component* dialog box, select the bearing as the *Selected Component*, the *Motion* method is *Distance*; in the copy option, select *Copy*. Select the bearing as the component to be copied; select the end face of the output shaft as the reference for the specified vector; enter 245 in the *Distance* text box, as shown in Figure 5-56; click the middle mouse button to complete the copy movement of the bearing, as shown in Figure 5-57.

(a) Move component parameter settings　　(b) Specified vector reference

Figure 5-56　Bearing movement setting　　Figure 5-57　Result of bearing movement

(9) Load sleeve: Select the *Add* button, open the *xingxing_youzhoutao* file under the folder *ch5\example*, and set the *Placement* method to *Constraint*. Select the constraint type as *Touch Align*, the *Orientation* method is *Preferred Touch*, and then select the sleeve end face and the bearing inner ring end face as references, as shown in Figure 5-58. Continue to select the constraint type as *Touch Align*, the *Orientation* method is *Infer Center/Axis*, and then select the cylindrical surface of the sleeve and the outer cylindrical surface of the bearing as references, and the system performs the axis touch constraint operation to complete the

assembly of the output shaft.

(a) Mating component reference   (b) Base component reference

Figure 5-58 Sleeve constraint settings

5.8 Questions and Exercises

1. Fill in the blanks

(1) The NX assembly module provides two basic assembly methods: _____and_____.

(2) In the component assembly, NX provides three parts to be placed and positioned, respectively_____, _____and_____.

(3) According to the number of constraints applied to the assembled part, it can be divided into four kinds of constraint states: _____, _____, _____and_____.

2. Choice

(1) _____in the assembly environment, it is moved or rotated within the degree of freedom of the component to facilitate establishing a constraint relationship between the components.

    A. Move component　　　　　　B. Edit component
    C. Constrain component　　　　　D. Suppres component

(2) Infer Center/Axis constraint needs to_____be specified.

    A. flat　　B. Reference axis　　C. Cylindrical surface　　D. Datum

(3) There can be_____work part in the NX assembly.

    A. four　　B. three　　C. two　　D. one

(4) In a circular pattern of components, you need to specify the_____and_____ parameters of the pattern.

    A. Axis, offset　　　　　　B. Direction, offset
    C. Axis, angle　　　　　　D. Direction, angle

3. Short answer questions

(1) What is a component?

(2) What are the occasions for bottom-up assembly and top-down assembly? Please briefly explain the steps for both assemblies.

(3) What is a reference set? What is the role of the reference set?

(4) Please list the types of the four assembly constraints.

(5) What are the characteristics of WAVE geometry linker technology?

(6) How to create an exploded view?

4. Exercises

Create a planet carrier assembly model:

This exercise is to create a planet carrier assembly model, as shown in Figure 5-59. The carrier model consists of a planet carrier, planetary shaft, planetary gear, internal ring gear and bearings. The planet carrier is provided with three planetary shaft holes, and the planetary shaft is mated with the hole; the planetary gear shaft is supported by bearing, and the inner ring of the bearing is mated with the planetary gear shaft; the planetary gear is mated with the planetary carrier through the bearing.

Figure 5-59　planetary carrier assembly model

Using bottom-up assembly method to create the assembly model. It mainly uses touch, align, inferred center/axis and other constraints for constrain assembly. It can be created by circular pattern when assembling the planetary gear and its shaft.

Chapter 6

Drafting

The engineering drafting is a graphical symbology that is produced according to certain rules on a two-dimensional plane or drawing and is used to express the design intent. The engineering drawing is the language of the engineer. The engineering drawings should accurately and completely express the structure, processing and assembly information of the mechanical parts. The engineering drawing solves the two-dimensional expression problems and also the three-dimensional shape problems by using the concept of projection. The three-dimensional structural features of the component can be converted into two-dimensional entities by projecting, so that people can express objective mechanical entities on a plane-like paper.

The Drafting application is designed to allow you to produce and maintain engineering drawings which comply with major national and international drafting standards. This chapter introduces the main functions of the NX drafting module and the common operations for creating 2D drawings based on 3D models.

The main contents of this chapter are:

(1) Understand the concept of the projection angle;

(2) To be proficient in the dependencies between engineering drawings and 3D solid models;

(3) Master the creation method of the drawing sheet;

(4) Master how to create standard-compliant views;

(5) Master how to create drafting dimensions, drafting annotations for part drawings;

(6) Master how to create assembly drawing and how to generate parts list.

6.1 Introduction to Drafting

Drawings can usually be divided into two categories: part drawing and assembly drawing. The part drawing can express the structure of the part and the tolerance annotations can also be used to express the size, shape, position, surface roughness and technical requirements of the part. In practice, these items are mainly used to guide the process planning and processing reference. An assembly drawing is a list of all the necessary components to complete the assembly, when the parts are assembled to form functional assembly or mechanisms, the assembly relationship between the parts (the associated dimensions between the geometric elements of the components, position requirements, etc.). In practice, it is used to guide the assembly and acceptance.

In order to fully and accurately express design information, early designers could only draw drawings by hand. From the perspective of function and importance, the engineering drawings at that time were irreplaceable. The original CAD software used computer graphics technology to realize the representation of two-dimensional lines on the computer, so that the functions of the pen and drawing paper can be detached and the graphic symbols can be drawn on the screen. But this technique only makes the designer get rid of the paper and the pen in form. From the essence of the design process, it is still a simple drawing of the two-dimensional line. With the continuous advancement of CAD technology, especially the development of solid geometry, the current CAD software has fully realized the three-dimensional modeling of mechanical components. Projected views of various directions and positions can be easily obtained using 3D model. This makes it unnecessary to express the structure through 2D drawings. However, considering the process planning and inspection requirements, the engineering diagram represents the function of processing and assembly information is still essential (this is mainly reflected in the annotation of the drawing). In any case, drawing work has become much easier due to advances in modeling.

Get started quickly with templates and examples that are easy to customize. The engineering drawing module of NX is mainly for the purpose of processing parts and manufacturing drawings. The projected view of the drawing can be created from an existing 3D model. Any 3D solid model created by the modeling module in NX can be used to generate a 2D drawing through the engineering drawing module. The generated drawing is completely associated with the solid model. As the solid model updates, the fully associative drafting annotations are automatically updated synchronously. Therefore, it can ensure the correctness of the two-dimensional engineering drawings and reduce the time required for 2D drawing updating due to changes in the 3D model.

6.2 Drafting Preferences

Before creating drawings, users should set the drawing standards and drawing habits, view preferences and annotation preferences for new drawings to ensure standardization of engineering drawings. You can control the size of the arrow, the thickness of the line, the size of the dimension font, and so on with the pre-setting of parameters. Once set, all newly created views and annotations will be consistent.

Custom drawing standards can be achieved through the following steps:

(1) Select menu *File > Utilities > Customer Defaults > Drafing > General/Setup > Standard*.

(2) Select a standard from the list of *Drafting Standard*. The optional standards are American standard ASME, German standard DIN, Russian standard ESKD, Chinese standard GB, international standard ISO and Japanese standard JIS.

(3) If the user wants to customize standard then click *Customize Standard*.

(4) Select any category in the list of *Drafting Standard* on the left side of the *Customize Drafting Standard* dialog box and modify any of the options on the tabbed page.

(5) Click *Save As*.

(6) In the *Standard Name* box, enter a new name.

(7) Click *OK* to save the custom standard.

After that, even if the computer is restarted, NX still guarantees the consistency of the customized drawing standards.

If users only use a specific view or annotation setting in the current drawing, it can be achieved by the following steps:

(1) Select the menu *Preferences > Drafting* to enter the drawing or annotation preferences.

(2) With drafting preferences and annotation preferences, the user can control specific parameters for this drafting. Workflow, drawing, and view options can be set by using the options available in the *Drafting Preferences* dialog box. With annotation preferences, users can change the dimension type, section lines, notes, labels, symbols, centerlines and hatches.

6.3 Drafting Management

Generating various projection views is the core of creating a drawing. In NX, any 3D model created by solid modeling can create multiple 2D drawings with different projection methods, different pattern sizes and different scales. The drawings created are all done by the drawing management function. NX drafing application provides management functions for various views, such as adding views, deleting views, aligning views, and editing views. These functions make it easy to manage the various types of views contained in the drawing, and to modify the scale, angle and other parameters between the views.

6.3.1 The 1st angle projection and the 3rd angle projection

Generally, there are two types of engineering projection methods: the first angle projection and the third angle projection. The main countries of the former are China, Britain, Germany and Russia. The latter mainly used in the United States, Japan, Singapore and other countries. These two projection methods have only different habits, there is no absolute advantage or disadvantage.

The three coordinate planes of the three-dimensional Cartesian coordinate system divide the space into eight quadrants. The first and third quadrants are respectively shown in Figure 6-1. The so-called 1st angle projection assumes that the part is placed in the first quadrant and the part is placed between the projection surface and the observer. Then, the parts seen by the observer are drawn on the projection surface that is the first quadrant angle projection view is formed. The so-called 3rd angle projection

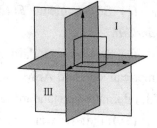

Figure 6-1 Cartesian coordinate system

assumes that the part is placed in the third quadrant and the projection surface is placed between the observer and the part. At this point, the observed part is drawn on the projection surface, which is the third angle projection view. For the 1st angle projection, the left view is on the right side of the front view, and the top view is below the front view. For a 3rd angle projection, the left view is on the left side of the front view, and the top view is on the top view, as if the part was placed in a transparent glass hexahedral box, as seen from the glass.

The difference between the 1st angle drawing and the 3rd angle drawing is the position where the view is placed. The 3rd angle drawing as shown in Figure 6-2: the left view is placed to the right while the right view is placed to the left, the top view is placed below, and so on. The 1st angle drawing, as shown in Figure 6-3: left view to the left while right view to the right, top view to the top, and so on.

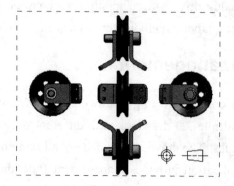
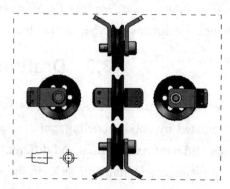

Figure 6-2 The 3rd projection view and symbol   Figure 6-3 The 1st projection view and symbol

The international standard stipulates that the engineering drawing can use the 1st angle projection method or the 3rd angle projection method and a special identification symbol is given separately.

6.3.2 Creating a new drawing

The user can create drawings using either a stand-alone workflow method or a model-based workflow method.

1. Stand-alone drawing workflow

The stand-alone drawing workflow is used to place the drawing data in a single part file. This is recommended for a 2D drafting process where only 2D geometry is used to create the drawing. The 2D curves can be placed directly on the drawing sheet, or can be placed in drawing views, and used to generate 3D model geometry.

2. Model-based drawing workflow

The model-based drawing workflow uses the existing 3D geometry to generate the 2D drafting data. In this workflow, user can:

Place a drawing directly in the file that contains the 3D model geometry. The drawing data is associated with the 3D geometry and will update when the model geometry updates.

Use the master model architecture, in which you place the drawing data in a file which is separate from the file that contains the model geometry. This workflow is recommended for a 3D drafting process. The drawing data is directly associated with the 3D model geometry and updates when the geometry updates, but different users can work concurrently with the same model data.

In the Drafting application, use one of the following methods to open the *Sheet* dialog box.

(1) Choose *Home*>*New Sheet* .

(2) Choose *Menu*>*Insert*>*Sheet*.

(3) In the *Part Navigator*, right-click the *Drawing* node and choose *Insert Sheet*.

The main options in the dialog box are as Table 6-1:

Table 6-1 The options of *Sheet* dialog box

Menu	Options	Description
Size	Use Template	Available when you create a new sheet. Makes a list of standard drawing templates available and adds the drawing border geometry to your part when the sheet is created
	Standard Size	Makes the size box available
	Custom Size	Lets you specify the height and length for a sheet
	Size	Available when you select Standard Size Lets you select a standard English or Metric drawing size from a list box
	Scale	Available only for the Standard Size and Custom Size options Lets you select the default view scale from a list, or set a specific default scale for all views added to the drawing Note: When modifying the Scale value while editing a sheet, only views on the current drawing which are not associated to expressions are affected

	Continued		
Menu	Options	Description	
Name	Sheets in Drawing	Lists all the sheets in the work part	
	Drawing Sheet Name	Sets the default name of the sheet, or lets you type a unique sheet name. You can type up to 30 characters for a name	
	Sheet Number	Drawing sheet numbers are comprised of an Initial Sheet Number (1), and if desired, an Initial Secondary Number (2), along with an optional Secondary Sheet Number Delimiter (3). Starting with the initial sheet character, each successive sheet is assigned in alphanumeric sequence. If a secondary character is desired, type it into the Initial Secondary Number box. This character must match the secondary character specified on the Sheet tab of the Drafting Preferences. The delimiting character separates the Initial Sheet Number from the Initial Secondary Number Initial, secondary, and delimiting characters appear in the Part Navigator under the Sheet Number column	
	Revision	Lets you type a unique revision letter for the new drawing sheet. The revision letter appears in the *Part Navigator* under the *Sheet Revision* column. To edit the revision letter of drawing, right-click its sheet node in the *Part Navigator* and choose *Increment Sheet Revision*. The next available revision letter is used	
Settings	Units	Specifies units for the sheet. If you change the unit of measure from Inch to Millimeters or vice versa, the *Size* options change to match the selected unit of measure	
	Projection	Specifies either 1st angle or 3rd angle projection. All projected and section views will display according to the set projection angle	

The *Size* option group includes three types of drawing creation methods:

(1) Use Template: Click the button to open the dialog box shown in Figure 6-4 and then users can directly select the system default template. The sheet template adds a sheet to the current part. Use the sheet template functionality to quickly add drawing sheets to the current work part with pre-defined content from the template. NX provides sets of drawing sheet templates in both English and metric sizes.

(2) Standard size: Click this button to open the dialog box as shown in Figure 6-5. In the *Size* drop-down list of the dialog box, select an appropriate standard size from A0-A4 as the sheet size of the current drawing. You can also directly select the scale of the drawing in the *Scale* drop-down list. *Sheet in Drawing* shows the name and number of all sheets included in the drawing.

(3) Custom size: Click this button to open the dialog box as shown in Figure 6-6. In this dialog box, you can customize the height and length of the new drawing in the *Height* and *Length* text boxes. In the *Scale* box, select the scale for the current drawing. This is suitable for situations where a non-standard frame is used.

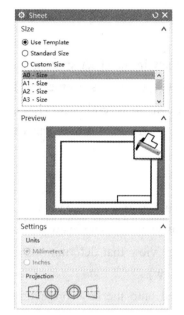  

Figure 6-4　Sheet template　　Figure 6-5　Standard size sheet　　Figure 6-6　Custom size sheet

6.3.3　Edit/Open/Delete sheet

There are usually three ways to edit a drawing sheet.

(1) Command Finder. *Edit Sheet* .

(2) Shortcut menu. Right-click the drawing border＞*Edit Sheet*.

(3) Part Navigator. Right-click the *Sheet* node＞*Edit Sheet*.

To open or delete a drawing sheet, we usually use *Part Navigator* to achieve. Double-click the *Sheet* node, and right-click the *Sheet* node＞*Open/Delete*, as shown in Figure 6-7 and Figure 6-8.

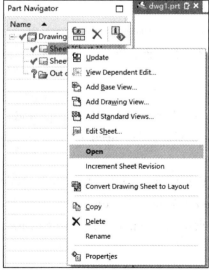 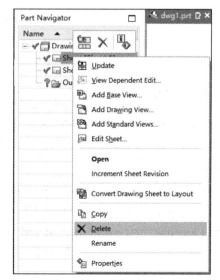

Figure 6-7　Open sheet　　　　　　Figure 6-8　Delete sheet

6.4 Drafting Views

Views are the basic elements that make up a drawing. The view in the sheet space is an integral or partial 2D section of the solid model. Adding a view operation is a process of generating each 2D section or view of the model. A drawing can contain several basic views, which can be the main view, projection view, section view, etc. A 3D solid model can be described by a combination of these views.

6.4.1 Base view

The first view placed on any sheet during the drawing process is called the base view. The base view can be used as a stand-alone view or as a parent view of other views. The parent view is an existing cartographic view, which is a reference view that determines the projection, alignment, and position of the newly added view (subview).

The base view is the graph from which the part is projected onto the base projection surface. It includes the front view, back view, top view, bottom view, left view, right view, isometric view and trimetric view. A drawing contains at least one basic view, so when generating a drawing, you should try to generate a basic view that reflects the main shape features of the solid model.

To create a basic view, you can open the *Base View* dialog box through the command finder, menu, graphics window or part navigator, as shown in Table 6-2 and Figure 6-9. Other views are opened in a similar way and will not be described later.

Table 6-2   Open the base view dialog box

Options	Operation
Command Finder	*Base View*
Menu	*Insert＞View＞Base*
Graphics Window	Right-click a drawing sheet border＞*Add Base View*
Part Navigator	Right-click a sheet node＞*Add Base View*

6.4.2 Projected view

In general, a single basic view is difficult to completely express the shape and structural features of a solid model. After adding the basic view, you need to add corresponding projection views.

After creating a basic view, continue dragging the mouse to add additional projection views of the base view as shown in Figure 6-10. With the *Projected View* dialog box, you can set the placement location, placement method and reverse view direction of the projected view.

Chapter 6 | Drafting | 157

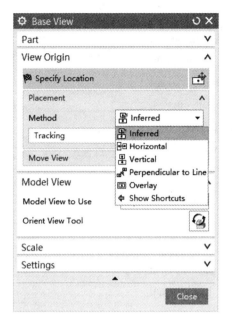 

Figure 6-9 *Base View* dialog box   Figure 6-10 *Projected View* dialog box

6.4.3 Detail view

A detail view is a portion of an existing view. The scale of the detail view may be adjusted independently of its parent view so that objects shown in it are more easily seen and annotated. You can create detail views with either circular or rectangular view boundaries, as shown in Figure 6-11. Associative view and scale labels can be attached to the detail view as well as to the detail boundary in the parent view.

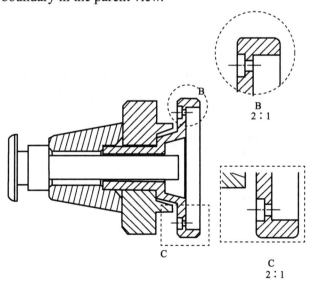

Fig 6-11 Circular and Rectangular detail views with label

Detail views are fully associative to their parent views. Any changes made to model geometry or view dependent edits made to the parent view are immediately reflected in the detail view. Like its parent view, you may need to update the detail view to clean up the display of hidden and silhouette edges. If the parent view contains 2D geometry such as sketches or curves, associative copies of them are placed in the detail view. These objects can only be edited in the parent view. Using the *Detail View* dialog box, as shown in Figure 6-12, users can set the type, boundary, parent view, scale, and annotation.

6.4.4 Section view

Section View is used to depict a part with a portion of it cut away in order to expose some or all of its internal features. The *Section View* command creates a true representation of the cut in the model and section profile shapes can be non-rectangular. The resultant geometry is

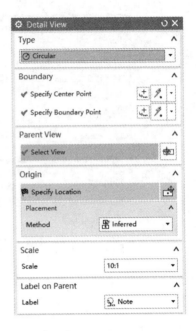

Figure 6-12 *Detail View* dialog box

persistent and is displayed in a new model view. These section views appear under the *Model Views* node in the *Part Navigator*. Note that the resultant geometry is added to your part, and the original model geometry is not affected.

Items that can be sectioned include: parts, assemblies, solids and sheets.

1. Simple section view

When the internal structure of the part is complex, the shape is relatively simple or the shape has been clearly expressed in other views. The internal structure of the part can be represented by a simple section view. To create a simple section view, you can open the *Section View* dialog box through the command finder, menu, graphics window and part navigator, as shown in Table 6-3. Other views are opened in a similar way, and will not be described later.

Table 6-3 Open the simple view dialog box

Options	Operation
Command Finder	*Section View* > *Simple/Stepped*
Menu	*Insert* > *View* > *Section View* > *Simple/Stepped*
Graphics Window	Right-click a basic view border > *Add Section View* > *Simple/Stepped*
Part Navigator	Right-click a basic view node > *Add Section View* > *Simple/Stepped*

The following is an example of the process of creating a simple section view.

(1) On the *Command Finder*, click *Section View*.

(2) Use the left mouse button MB1 to select the parent view (You can also right-click the border of the parent view and choose *Add Section View*).

(3) In the *Section View* dialog box, in the *Section line* group, set *Method* to *Simple/Stepped*, as shown in Figure 6-13.

(4) Move the dynamic section line to the cut position point (Turn snap point methods on or off to assist you in picking a point on the view geometry).

(5) Select the point to place the section line symbol, as shown in Figure 6-14.

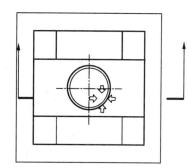

Figure 6-13 *Section View* dialog box   Figure 6-14 Select the cut position point (Hole center)

(6) Drag the cursor outside the view until the view is positioned correctly, as shown in Figure 6-15.

(7) Click to place the section view, as shown in Figure 6-16.

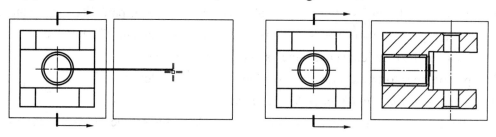

Figure 6-15 Section view position   Figure 6-16 Parent view (left) and simple section view (right)

2. Half section view

When the structure of the part is close to symmetry, it can be represented by a half section view. The half section view is a section view with half of the part sectioned and the other half not sectioned.

The specific steps to create a half section view are as follows:

(1) On the *Command Finder*, click *Section View* .

(2) In the *Section View* dialog box, in the *Section Line* group, set the *Method* to *Half* .

(3) Select the parent view, as shown in Figure 6-17.

(4) Use the *Snap Point* option *Arc Center* to locate the cut position, as shown in Figure 6-18.

(5) Select a second point to place the bend segment, as shown in Figure 6-19.

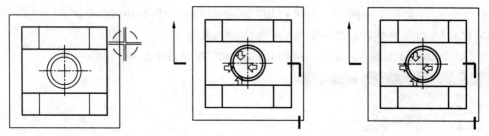

Figure 6-17　Select parent view　　Figure 6-18　Cut position　　Figure 6-19　Second point of the bend

(6) Drag the cursor to orient the section line symbol in the parent view, as shown in Figure 6-20.

(7) Click to place the view, as shown in Figure 6-21.

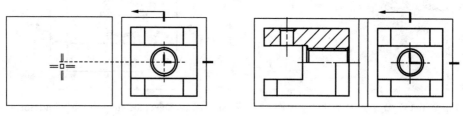

Figure 6-20　Orient the section line symbol　　　　Figure 6-21　Place the view

3. Revolved section view

In a revolved section view, the section line symbol consists of two legs revolving about a common rotation point, usually at the axis of a cylindrical or conical part. Each leg contains one or more cut segments, connected to each by a circular bend segment. The revolved section view unfolds all of the individual cut segments onto a common plane.

The specific steps to create a revolved section view are as follows:

(1) On the *Command Finder*, click *Section View* .

(2) In the *Section View* dialog box, in the *Section Line* group, set the *Method* to *Revolved* .

(3) Select the parent view, as shown in Figure 6-22.

(4) Select a rotation point to place the section line symbol, as shown in Figure 6-23.

(5) Select a point for the first segment, as shown in Figure 6-24.

 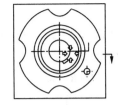 

Figure 6-22 Select parent view   Figure 6-23 Select a rotation point   Figure 6-24 Point for the first segment

(6) Select a second segment point, as shown in Figure 6-25.

(7) Next, we add an additional segment to the second leg of the section line.

(8) In the background of the graphics window, right-click and choose *Specify Leg 2 Location*.

(9) Select a point to define the new segment, as shown in Figure 6-26.

(10) Click and drag the section line handles to refine the placement of the cut segments, bend segment, origin point, and arrow locations, as shown in Figure 6-27.

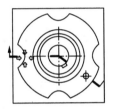 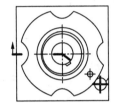

Figure 6-25 Second segment point   Figure 6-26 Select a new line   Figure6-27 Define new segment

(11) Right-click and choose *View Origin*, and then drag the view to an appropriate location and click to place it, as shown in Figure 6-28.

4. Stepped section view

A stepped section view consists of multiple cut segments passing through the part. All cut segments are parallel to the hinge line and are attached to each other with one or more bend segments.

The specific steps to create a stepped section view are as follows:

(1) On the *Command Finder*, click *Section View* 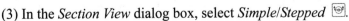.

(2) Use the left mouse button MB1 to select the parent view (You can also right-click the border of the parent view and choose *Add Section View*).

(3) In the *Section View* dialog box, select *Simple/Stepped* 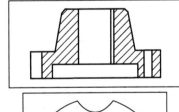.

(4) Turn snap point methods on or off to assist you in picking a point on the view

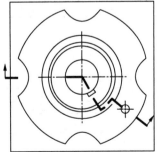

Figure 6-28 Place the revolved section view

geometry.

(5) Move the dynamic section line to the desired cut position, as shown in Figure 6-29.

(6) Orient the cut direction as desired, then right-click and choose *Align* to *Hinge* to lock the orientation of the section view.

(7) Right-click and choose *Section Line Segments*, as shown in Figure 6-30.

(8) Select the next point to place a cut segment.

(9) Select additional points to add subsequent cut segments as desired, as shown in Figure 6-31.

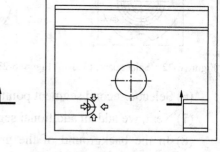

Figure 6-29　Select cut position

(10) In the background of the graphics window, right-click and choose *View Origin*, and then move the cursor to the desired location, as shown in Figure 6-32.

(11) Click to place the view, as shown in Figure 6-33.

Figure 6-30　Section line segments

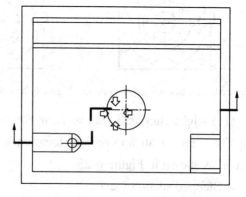

Figure 6-31　Place cut segment

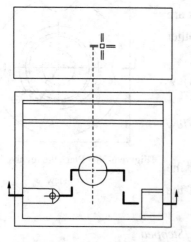

Figure 6-32　Place stepped view

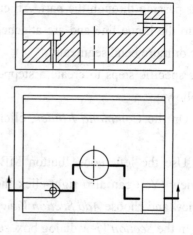

Figure 6-33　Stepped sction view

6.4.5 View break

For some long-axis parts or complex parts with large shape and size, in order to make the view easy to express and save the view space, you can use the *View Break* command 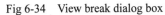 to add multiple horizontal or vertical view breaks, as shown in Figure 6-34.

You can create the required section by setting the offset distance of the section line and determining the position of the two sections in the parent view, as shown in Figure 6-35.

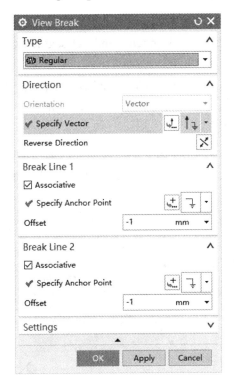

Fig 6-34　View break dialog box

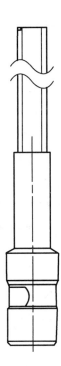

Fig 6-35　Regular view breaks

6.5　Edit Drafting View

The editing of view objects mainly includes two categories: a single view as an edit object and a specific line in the view as an edit object. The former includes graphics style editing, section line editing, view deleting; the latter includes adding, deleting and moving lines.

6.5.1　Setting dialog box

Use the Settings dialog box to edit the appearance and general properties of a drafting view. Use any of the following methods to display the *Settings* dialog box:

- Double-click a view boundary.

- Right-click one or more view boundaries and choose *Settings* from the shortcut menu, or click the Settings button on the shortcut toolbar.
- Click the *Edit Settings* command and then select one or more views from the drawing sheet.
- Highlight the view border, press and hold the right mouse button, then select *Settings* from the radial toolbar.

6.5.2 Section line and hinge line

A section line symbol consists of arrow segments, bend segments and cut segments, as shown in Figure 6-36.

A hinge line is a linear reference used to orient the section view cut position(s) in the parent view. When creating simple, stepped, or half section views, the section line cut segments are parallel to the hinge line.

When you define the hinge line, you can choose whether or not to create an associative link between the hinge line and the model geometry used to define the orientation of the hinge line, as shown in Figure 6-37.

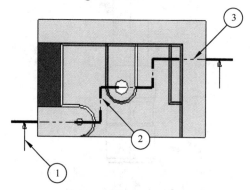
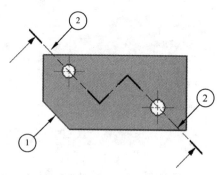

Figure 6-36 Section line symbol
1—arrow segment ; 2—end segment; 3—cut segment

Figure 6-37 Hinge line and cut segment
1—hinge line; 2—cut segment

6.5.3 Non-sectioned component

Chinese drawing standards stipulate for bolts, pins, shafts, ribs and other parts or geometries, even if the cutting plane passes through them, usually do not cut, then how to achieve in NX?

For non-sectioned component, you can use the *Section in View* dialog box to achieve, as shown in Figure 6-38.

Select the non-sectioned view on the View option as shown in Figure 6-39. Then in the Body or Component option, select the non-sectioned component as shown in Figure 6-40.

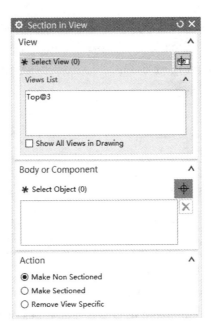

Figure 6-38 Section in view dialog box

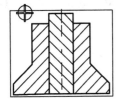 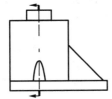 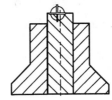 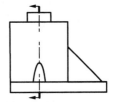

Figure 6-39 Select non-sectioned view   Figure 6-40 Select non-sectioned component

Select *Make Non Sectioned* in the action options, click *OK*. You need to click the right mouse button on the view boundary to select the *Update* view, as shown in Figure 6-41.

6.6 Drafting Annotations

After completing the creation and placement of the view, it is usually necessary to label and annotate. For most engineering drawings, the following three types of labels are mainly included.

6.6.1 Centerlines

If you don't make any special settings, NX automatically creates a centerline for features such as holes or pins in any existing view. For features such as slots, the centerline is not automatically created. For a circular array of holes, just create a linear centerline for each hole. This requires the user to manually relabel the centerline that does not meet the design intent.

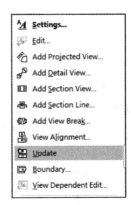

Figure 6-41 Update view

Use the *Center Mark* command to create center marks through points or arcs, as shown in Figure 6-42. A center mark that passes through a single point or arc is called a simple center mark, as shown in Figure 6-43.

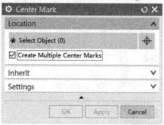

Figure 6-42　*Center Mark* dialog box　　　Figure 6-43　Simple center mark

Use *Circular Centerline* to create full or partial circular centerlines through points or arcs. The radius of the circular centerline is always equal to the distance from the center of the circular centerline to the first point selected(Figure6-44).

Use the *2D Centerline* command to create centerlines between two edges, two curves or two points. You can create 2D centerlines using curves or control points for limiting the length of the centerline. For example, if you use control points to define the centerline (from arc center to arc center), a linear centerline is produced(Figure6-45).

 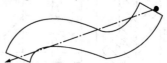

Figure 6-44　2D Centerline　　　Figure 6-45　Linear Centerline

6.6.2　Note

Use *Note* to create and edit notes and labels. A note consists of text, while a label consists of text with one or more leader lines. Text can be imported by reference to expressions, part attributes, object attributes and can include symbols formed from control character sequences or user-defined symbols.

On Windows, while in the *Drafting* application you can drag a text file onto a drawing sheet to create a note.

(1) Open the *Note* dialog box in one of these ways:
- Choose *Home* tab＞*Annotation* group＞*Note*.
- Choose *Menu*＞*Insert*＞*Annotation*＞*Note*.

(2) Type the desired text in the *Text Input* box, as shown in Figure 6-46. Text appears in the text box and at your cursor location in the graphics window.

Figure 6-46　*Note* dialog box

(3) Move the cursor to the desired location and click to place the note.

6.6.3 Geometric dimension and tolerancing in drafting

Geometric Dimensioning and Tolerancing (GD&T) is a system for defining and communicating engineering tolerances. It uses a symbolic language on engineering drawings and computer-generated three-dimensional solid models that explicitly describes nominal geometry and its allowable variation. It tells the manufacturing staff and machines what degree of accuracy and precision is needed on each controlled feature of the part. GD&T is used to define the nominal (theoretically perfect) geometry of parts and assemblies, to define the allowable variation in form and possible size of individual features, and to define the allowable variation between features. Geometric dimensioning and tolerancing is an engineering language. It allows manufacturing and production units to more clearly understand design requirements. The geometric tolerancing functions manipulate actual design level manufacturing data and allow a designer to include manufacturing information in the form of tolerance features directly in the master model of the part. These tolerance features (datums, datum targets, and tolerances) are "smart", knowing when a change in the model influences a change in a tolerance feature. Simple checking is done automatically, and comprehensive checking is done on user demand.

1. Drawing dimensions

Dimensions are used to identify the size of the geometric features on the solid and the tolerances allowed by the machining. Since NX drafting and NX modeling are fully related, once the solid model changes the dimensions marked on the drawing will change. Therefore, if you want to change the size parameters in the part, it is best to modify it in the 3D solid model and the corresponding size in the drawing will be updated automatically which ensures the consistency of the drawing and the model.

NX provides 19 dimension types for dimensioning on part views, and Table 6-4 lists several common dimension types.

Table 6-4 Dimension Types

Types	Meaning	Symbole
Horizontal	Create dimensions that are measured parallel to the X axis	
Vertical	Create dimensions that are measured parallel to the Y axis	
Point–to–Point	Create the shortest distance dimension between the two point	
Perpendicular	Create a vertical dimension between a line or centerline and also between points	
Angular	Create an angular dimension between two non-parallel lines	
Cylindrical	Creates a cylindrical dimension that is a linear distance between two objects or points that measures the outline view size of the cylinder	
Diametral	The diameter dimension is marked for any circular feature by a single leader, and the dimension text contains a diameter symbol	

2. Geometric tolerance

The Feature Control Frame dialog box provided by NX can be used to mark the geometrical tolerance symbols on the part drawing. It can create single and compound geometric tolerance symbols. Often associated with datum, use Geometric tolerance to create and place basic dimension notations in the drawing.

The following example creates a feature control frame from the Drafting application.

(1) Choose *Home* tab＞*Annotation* group＞*Feature Control Frame*, as shown in Figure 6-47.

(2) In *Frame* group, set *Characteristic* to *Straightness* and set *Frame Style* to *Single Frame*.

(3) In *Tolerance* options, set the tolerance shape to *Diameter* and set 0.1 in the tolerance value box.

(4) In *Text* options, set the text to display above the tolerance grid.

(5) Click to place the symbol, as shown in Figure 6-48.

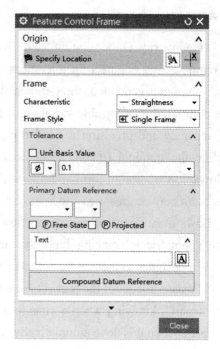

Figure 6-47　Feature control frame

The simplified method of compound geometric tolerance can drag and drop different geometric tolerance grids onto the existing grid until the dotted frame appears, click the mouse as shown in Figure 6-49.

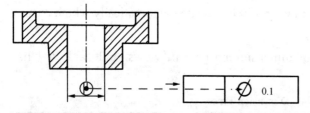

Figure 6-48　Add a single geometric tolerance

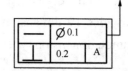

Figure 6-49　Compound geometric tolerance

3. Datum

A datum is theoretical exact plane, axis or point location that GD&T or dimensional tolerances are referenced to. You can think of them as an anchor for the entire part; where the other features are referenced from. A datum feature is usually an important functional feature that needs to be controlled during measurement as well. All GD&T symbols except for the form tolerances (straightness, flatness, circularity and cylindricity) can use datums to help specify what geometrical control is needed on the part. When it comes to GD&T, datum symbols are your starting points where all other features are referenced from. A datum is a theoretical, perfect point, axis, or plane that establishes a stable and immobile origin of measurement for tolerances. A datum definition derives a datum from a high-quality existing point, plane, or axis inspection feature, or if the datum does not lie on a part or is composed of several features, from a constructed feature that references such a high-quality feature. Since datum definitions do not produce actual values, but act instead as references that never vary geometrically, a datum is not itself a tolerance.

You can create a datum feature symbol by following steps:

(1) Choose *PMI* tab > *Annotation* group > *Datum Feature Symbol* , as shown in Figure 6-50.

(2) In the *Datum Feature Symbol* dialog box, specify any automatic *Alignment* options that you want.

Figure 6-50 Datum Feature Symbol

(3) Use the *Orientation* options to define the annotation plane of the datum feature symbol.

(4) If you want to change the default datum feature symbol letter, enter the new letter or series of letters in the *Letter* box.

(5) Click to open the *Settings* dialog box if you want to specify styles for the appearance of the datum feature symbol.

(6) When you are ready to create the datum feature symbol, drag the symbol to specify its location, and click to place it.

4. Surface finish

The *Surface Finish* command is used to create standards-compliant surface finish symbols on drawing to designate surface textures of solid materials.

This example procedure shows how to add an ANSI standard surface finish symbol with a leader, and to associate the symbol with a surface.

Choose *Annotation* group > *Surface Finish* .

(1) In the *Leader* group, make sure *Type* is set to *Plain* and *Arrowhead* is set to *Filled Arrow*.

(2) In the *Attributes* group, make sure the *Standard* is set to ANSI.

(3) From the *Material Removal* list, select *Open, Modifier*.

(4) From the *Roughness* list, select *Ra value*.

(5) Click on the edge of the part and drag to place the surface roughness symbol with the leader, as shown in Figure 6-51.

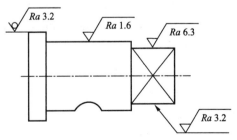

Figure 6-51 Surface finishing symbol

6.6.4 Assembly drawing

In assembly drawings, you can display the parameters of assembly features and all assembly components; however, you cannot show the dimensions of subassemblies. The dimensions in an assembly drawing are visible only at the assembly level. You must have the assembly from which a drawing was created in session in order for the dimensions to appear for modification. Like single part files, you can create your drawing directly in your assembly file, or add the assembly as a master part to a non-master drawing file. Once the drawing is created, you can use the drafting tools to create and annotate views on separate drawing sheets.

The Parts lists are derived directly from the components in the assembly, regardless of how the drawing is created (in the master part or in a non-master part). Dimensions, labels, symbols, and other drafting aids are fully associative with the geometry in the component to which they are attached. Similar, drawing-dependent modifications made at the drawing level are retained when the component geometry is edited.

1. Assembly dimension

Assembly dimension needs to add the suffix text by using the *Appended Text* dialog box to express the assembly type.

Select the two edges of the dimension, as shown in Figure 6-52. Right-click and open the *Appended Text* dialog box, as shown in Figure 6-53. In the *Controls* group, from *Text Location* list, select *After*. In the *Text Input* group, enter assembly dimension.

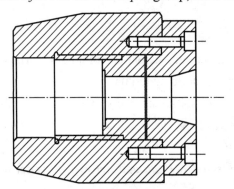
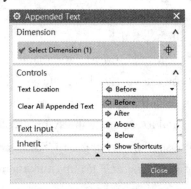

Figure 6-52   Select the drafting dimension   Figure 6-53   Appended text dialog box

Click in the appropriate position to place the assembly dimension, as shown in Figure 6-54.

2. Parts list

Parts lists are derived directly from the components listed in the Assembly Navigator, thus they provide an easy way for you to create a bill of materials (BOM) for your assembly. Because parts list is a unique form of a table, all the interactions used to manage the contents of a table are also used to manage the contents of a parts list.

NX can automatically generate parts lists and part numbers for assembly parts. The premise is that the user must add at least the DB_PART_NO attribute for each type of part model, and set the corresponding string value, as shown in Figure 6-55.

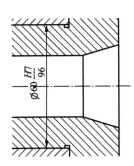 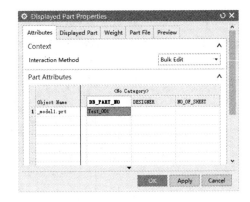

Figure 6-54   Add an assembly dimension   Figure 6-55   Add part attribute

NX adds China National Standard, select *Utilities＞User Defaults＞Drafting＞General＞Standard*, as shown in Figure 6-56.

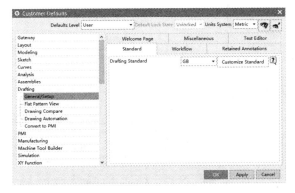

Figure 6-56   Set China National Drafting Standard

After completing the above settings, you can click *New*, and then on the *Drawings* tab of the *New* dialog box, select the assembly template, and provide the assembly model name, as shown in Figure 6-57.

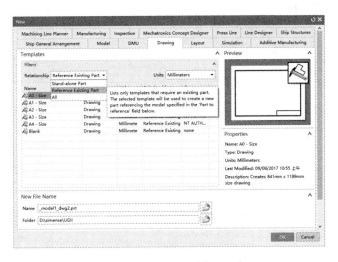

Fig 6-57   Create assembly template

Click OK to automatically generate a parts list. After setting the views on the drawing sheet, select the parts list in the navigator and right click. Select Auto Balloon to automatically generate the part number, as shown in Figure 6-58 and Figure 6-59.

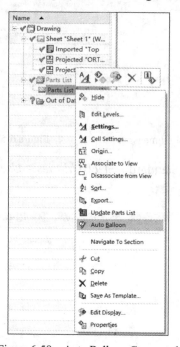

Figure 6-58   Auto Balloon Command

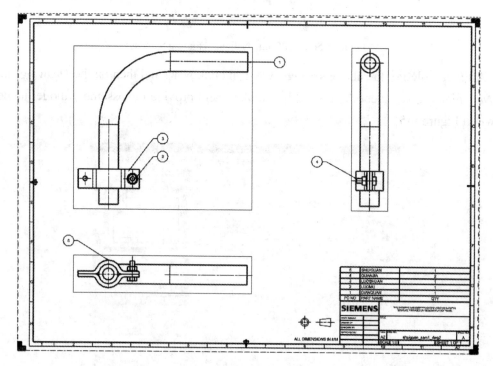

Figure 6-59   Create parts list

6.7 Exercise-Drafting Example

Create an engineering drawing of the part shown in Figure 6-60.

(1) Select the A2 drawing based on the part size. Use the *Populate Title Block* command to edit the contents of individual unlocked title block cells (Figure 6-61). You can also right-click the title block and use the *Populate* command.

Figure 6-60   Part model   Figure 6-61   *Populate Title Block* dialog box

(2) Select a 1:5 scale and use the GB standard template.

(3) Click *Start＞Drawing＞New*, open *Sheet* dialog box.

(4) In the *Size* group, select *Use Template＞A2-Size*, as shown in Figure 6-62.

(5) Add basic view by *Base View* command, as shown in Figure 6-63. You can use *Orient View* command to adjust view orientation, as shown in Figure 6-64. Arrange the desired view in the right place on the drawing, as shown in Figure 6-65.

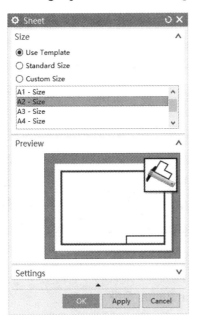

Figure 6-62   Select template   Figure 6-63   Add base view   Figure 6-64   Orient View

(6) Add projection view. Click *Projected View* command, and open dialog box. Select

base view as "parent view". The system automatically generates a projection view based on the "parent view", move the mouse to the appropriate position under the "parent view", click the mouse to place a top view, as show in Figure 6-66.

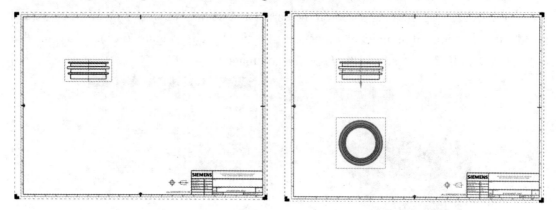

Figure 6-65　Add base view　　　　　　Figure 6-66　Add projection view

(7) Add section view. Click *Section View* command, and select the base view as "parent view". Select the middle hole as the cutting position, and drag the mouse to the appropriate position to give a section view, as shown in Figure 6-67.

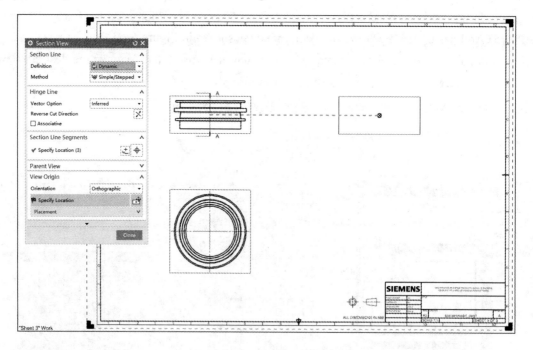

Figure 6-67　Add section view

(8) Create geometric dimension and tolerance, as shown in Figure 6-68.

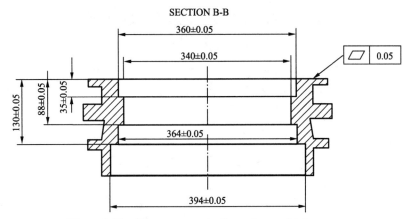

Figure 6-68 Create geometric dimensions and tolerances

(9) Add technical requirements, as shown in Figure 6-69.

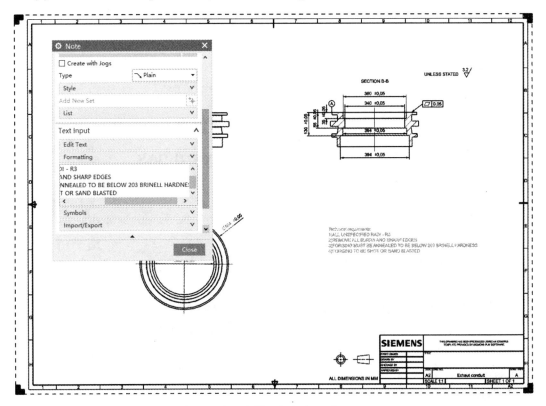

Figure 6-69 Add technical requirements

(10) At last, add part attribute to automatically generate parts list when generating assembly drawings later. Click *Properties* on the *File* menu, add the "DB_PART_NO" title to the part drawing, set the numbered text, and add "DB_PART_NAME", and set the part name to "*Exhaut conduit*", as shown in Figure 6-70.

Figure 6-70　Add technical requirements

6.8　Questions and Exercises

1. Fill in the blanks

(1) With＿＿＿＿and＿＿＿＿, the user can control the default behavior of a specific parameter for a single charting application.

(2) Any 3D model created using solid modeling can create＿＿＿＿2D drawings with different projection methods, sizes and scales.

(3) There are two common engineering projection methods:＿＿＿＿and＿＿＿＿.

(4) ＿＿＿＿is the most basic element that makes up a drawing.

(5) For parts like a bolt, pin, shaft, etc. on the assembly drawing, you can use the＿＿＿＿dialog box on the View Editing toolbar to achieve non-sectioned.

2. Choice

(1) For a model or assembly file, you can include both＿＿＿drawings.

　　A．1　　　　　　B．2　　　　　　C．3　　　　　　D．multiples

(2) Which one is right?＿＿＿

　　A．Drawing presets are not available in the modeling application.

　　B．A partial section view is a direct cut on the original view instead of generating a new section view.

　　C．Only one drawing can be generated for one part.

D. Once the various parameters of the 2drawing are set, they cannot be changed.

(3) The section is partially cut with a cut surface, and the resulting section view is called _____.

    A. Break-out section view   B. Revolved view

    C. Detail view   D. Half-section view

3. Short answer questions

(1) Briefly describe the similarities and differences between the 1st angular projection and the 3rd angular projection.

(2) Briefly describe the key steps to create a partial section view.

(3) How to deal with uncut parts and describe the key steps.

(4) Briefly describe how to generate the parts list and part number on the assembly drawing.

4. Exercises

(1) Open ch6\exercise\daxiang_ti.prt, as shown in Figure 6-70. Given the part model, it is required to use different views to accurately represent the structure of this part.

(2) Open ch6\exercise\jian_shu_qi.prt, as shown in Figure 6-71. Given the reducer model, it is required to mark the necessary dimensions, part number and parts list.

Figure 6-70   Part model   Figure 6-71   Reducer model

Chapter 7
Product and Manufacturing Information

Product and Manufacturing Information (PMI) is information that you attach to or associate to a part or assembly in 3D environments such as Gateway and Modeling. One of the benefits of PMI is that when added to a 3D model, it can be used by downstream applications for manufacturing, assembly planning, analysis, inspection and collaboration. PMI created with NX complies with industry standards for 3D digital product definition, and 3D models with PMI can be used as a replacement for drawings. You can display PMI information in any model views in the part file and at any level in an assembly structure. You can use assembly filters to control which PMI information contained in component parts is displayed in an assembly.

This chapter mainly introduces configuration, dimension, annotation, and section view in PMI.

After completing this chapter, you will understand:

(1) Regulation and consideration of PMI;
(2) Characteristics of PMI dimension and PMI annotation;
(3) Operating methods of PMI section view;
(4) Process for how to add MPI to a 3D model1.

7.1 Configuration

Before annotating the 3D digital model with NX PMI module, the PMI module should be set up in advance. Use the PMI Preferences command to control the default behavior of PMI-specific functionality. You can use the command options to:

- Set the default annotation plane for PMI.
- Specify the visibility and readability of PMI in model views.

- Control the size of PMI in the display.
- Control the orientation and position of PMI in the screen display.
- Set the initial letter for datum labels.
- Set the default size of rectangular, circular, and annular PMI regions.
- Set crosshatching and associative plane options for section views.
- Control display of a dialog box for Find PMI Associated to Geometry when that command is invoked from a shortcut menu.
- Set properties for a section view cutting plane symbol.

Select *Menu File> Preferences > PMI*. The setting of *PMI Preferences* is shown in Figure 7-1.

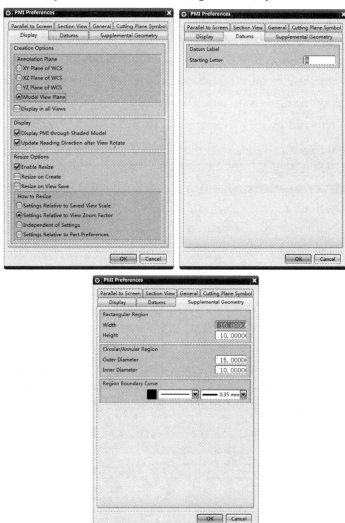

Figure 7-1　Setting of PMI Preferences

In the *Display* tab, *Annotation Plane* refers to the annotation plane of a 3D digital model. The XY Plane of the local coordinate system, the XZ Plane and the YZ Plane can be used as the annotated plane. Multiple local coordinate systems can be set in the model, so it can be labeled on multiple parallel XY planes, XZ planes and YZ planes, which enlarges the scope of labeling. The Model View Plane refers to the plane in which the computer display screen is located. The setting annotation plane here refers to the default annotation plane when the annotation is created. During the annotation process, the required annotation plane can be selected according to the needs of the annotation. In the *Display* option, tick the *Update Reading Direction* after *View Rotation* to ensure that the dimensions are correctly displayed. The initials of the datum symbol are set in the *Datums* tab. The size range of the replenishment geometry can be set in the *Supplemental Geometry* tab.

In addition to setting up the PMI preferences, you need to set up the annotation style. In the *Drafting Preferences* dialog box, the dimension and annotation information can be preliminarily set. The settings should meet the requirements of national standards. Linear dimension values should generally be marked above the dimension line and allowed to be marked at the break of the dimension line. The dimension of the unit is selected by millimeter, and the dimension type can be selected according to the need of marking. For dimension lines, solid arrows, slants, or dots can be used to mark them, depending on the labeling situation. The settings are shown in Figure 7-2.

Figure 7-2　Settings of dimension and text

7.2 Creating PMI

7.2.1 PMI annotation plane

PMI annotation and dimension display instances are placed on an annotation plane in 3D space. NX tries to determine an appropriate orientation and location for PMI display instances from selected geometry. If the selected geometry does not completely define the annotation plane, or if the PMI is not associated with or attached to any object, NX uses information specified in the *Orientation* group of the *PMI* dialog box as a guide for determining the annotation plane.

Select Any *PMI* dialog box > *Origin* group > *Plane* list. Plane options include:
- XY Plane - Parallel to the XY plane of the WCS.
- XZ Plane - Parallel to the XZ plane of the WCS.
- YZ Plane - Parallel to the YZ plane of the WCS.
- Model View Based - Perpendicular to the model view line of sight.
- Last User Defined - The last user defined plane.
- User Defined - Exactly the plane you specify.

A user-defined annotation plane tracks the plane you specify. If the specified plane moves, the annotation plane of any PMI object using the user-defined plane changes. If geometry that is used to define an annotation plane is removed (is deleted, suppressed, changes type, or is unloaded), NX continues to display PMI instances on the defined plane. PMI may or may not become retained, as described in the following cases. If NX infers the annotation plane of a PMI display instance based on selected geometry, and that geometry is removed or deleted, the PMI is placed in a retained state. If the annotation plane for the PMI display instance is user-defined, the PMI is not placed in a retained state unless the geometries to which the PMI leader or dimension extension line is attached, is removed or deleted. When the orientation of the annotation plane is one of the principal planes of the WCS, the planes of the *OrientXpress* tool are available in the graphics window. You can click a tool plane to reset the annotation plane. When no dialog boxes or commands are active, you can drag PMI objects to a new location on their annotation plane. PMI leader arrowheads are always drawn in the same plane as the terminating leg of the leader line. If the termination point for the leader does not lie on the annotation plane, arrowheads may appear foreshortened in the annotation plane view.

7.2.2 PMI associated objects

The Associated Objects option lets you explicitly capture your design intent by directly specifying the portions of the model to which the PMI pertains. This eliminates any confusion or question about which objects in the model are affected by the PMI. Select Any *PMI* dialog box>*Associated Objects* group>*Select Object* (Figure 7-3). Selectable objects vary depending on the PMI type, but may include points, edges, faces, bodies, routing objects and components.

Figure 7-3　Associated objects

If you have a note that specifies a particular finish for several faces, but you only want a single note leader pointing to one of the faces, you can attach the leader to that face and use the *Associated Objects* option to associate all the faces to the note. Whenever the note is queried or selected, its associated faces highlight in the graphics window and are listed in the query results.

7.2.3　Sizing PMI

The initial size of PMI objects, such as text, arrows, lines and so forth, depends on the values that are specified for the different annotation objects in the *Drafting Preferences* dialog box Figure7-4. In addition, the preferences in the *PMI Preferences* dialog box let you control how and when PMI is resized in the model view. Table7-1 provides a quick reference for how to control the size of your PMI objects in a model view. (To resize PMI, the *Enable Resize* check box in the *PMI Preferences* dialog box must be selected.)

Figure 7-4　Sizing PMI

Table 7-1 Quick reference for how to control the size

To	Do this
Adjust the default size of existing PMI.	(1) Right-click one or more PMI objects and choose *Edit Settings* (2) In the *Settings* dialog box, change the values that control the size and appearance of the PMI objects (Figure 7-4)
Resize PMI in the current view, and in any other view in which the PMI is displayed with respect to the scale or zoom factor of the view.	In the *PMI Preferences* dialog box, on the Display tab, in the *How to Resize* group, click an option. Choose *PMI* tab>*Resize PMI*
Reset PMI in the current view, and in any other view in which the PMI is displayed, to the current annotation preferences specified in the *Drafting Preferences* dialog box.	In the *PMI Preferences* dialog box, in the *How to Resize* group, select the *Settings Relative* to *Part Preferences* check box. Choose *PMI* tab>*Resize PMI*
Reset all PMI in the work part to the current annotation preferences specified in the *Drafting Preferences* dialog box.	Choose *PMI* tab→ *Reset All PMI Settings*

7.3 Dimensions

NX supports the creation of standard dimension types, including baseline, chain and coordinate sizes. Using dimension options, you can create and edit various dimensions, and set local preferences to control the display of size types. NX predicts and creates dimensions based on the selected objects using the intelligent judgment algorithm. If the model contains product and manufacturing information (PMI) features, you can use the inheritance PMI option of the drawing view to inherit the PMI annotations and dimensions directly from the model and place them on the drawing. In addition, size of hole can be automatically generated from the feature parameters used to create feature of hole using the hole annotation options in both linear and fast commands.

7.3.1 PMI dimensions

PMI dimensions can be created and displayed for models in any chosen orientation. PMI dimensions are similar to drafting dimensions, but they are calculated and displayed in 3D space on an annotation plane. Therefore, you must carefully consider the annotation plane when you create PMI dimensions.

If the model contains a sphere, the dimensions and guide lines should be included in the marking plane containing the center of the sphere. If there is a cylindrical surface in the design model, the dimensions and guide lines should be marked on the axis perpendicular to the cylindrical surface or in the marking plane containing the cylindrical axis. If the design model contains two parallel planes, the dimension and the guide line should be marked on the marking plane perpendicular to the feature center plane and the distance between the two parallel planes. The dimension boundary line is drawn with a thin solid line, and should be drawn from the outline, axis or symmetrical centerline of the graph. The contour line, axis or symmetrical centerline can also be used as the dimension boundary line. The dimension boundary line of the marking angle should be drawn along the radial direction, the dimension boundary of the marking chord length should be parallel to the vertical bisector of the chord, and the dimension boundary of the marking arc length should be parallel to the angular bisector of the arc to the center of the circle. When the radian is large, it can be drawn along the radial direction.

The dimension line is drawn with a thin solid line. The terminal of the dimension line can be either an arrow or a slash. Arrows are commonly used in mechanical drawings. When dimension lines and dimension boundaries are perpendicular to each other, only one dimension line terminal can be used in the same drawing. When dimensioning a linear dimension, the dimension line should be parallel to the marked line. The terminals of the diameter of the arc and the radius of the arc should be painted into arrows. When dimensioning an angle, the dimension line should be drawn into an arc, the center of which is the vertex of the angle. When marking the diameter, the ϕ should be added before the size number. When marking the radius, the R should be added before the size number. When marking the radius or diameter of the sphere, the S should be added before the ϕ or R. When the reference size is marked, the numbers should be added with brackets. Linear dimension values should generally be marked above the dimension line, but also allowed to be marked at the break of the dimension line, and parallel to the dimension line. For narrow sizes, dots or slashes are allowed to replace arrows if there are not enough places to draw arrows or write size numbers.

Select *PMI* tab>*Dimension* group>select any dimension type(Figure7-5). The *PMI* dimensions types include *Rapid Dimension*, *Linear Dimension*, *Radial Dimension*, *Angular Dimension*, *Chamfer Dimension*, *Thickness Dimension*, *Arc Length Dimension* and *Ordinate Dimension*.

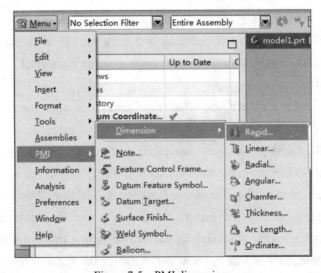

Figure 7-5 PMI dimensions

7.3.2 Converting feature dimensions to PMI

You can convert a feature dimension to a PMI feature dimension using the Display as *PMI* option in the *Feature Dimension* dialog box. Select *Menu File > Edit > Feature > Feature Dimension > PMI* group *> Display as PMI* (Figure 7-6). A PMI feature dimension has the following behavior:

(1) It is associative to the feature.

(2) It supports bidirectional editing. Changes you make to the PMI feature dimension are applied to your model, and viceversa.

If you make changes to PMI in a non-master part, you must have write access to the master model containing the PMI, and the master model must be loaded into your NX session.

Figure 7-6 Converting feature dimensions to PMI

7.3.3 Tips for using dimensions

These methods can speed up the creation and management of dimensions, the tips as shown in Table 7-2.

Table 7-2 Tips for using dimensions

Objective	Operation
Display control panel display input box when creating and editing dimensions	(1) Adjust the suspended screen display dialog box delay mapping preferences. This option controls the time when the cursor is needed before the control panel appears (2) If you want to temporarily hide the screen input box, press <F3>. Press <F3> again to re display the display input box
Temporarily suppress all automatic alignment control (such as horizontal or vertical alignment and automatic center placement).	Hold down the <Alt> key while placing the dimension
Move the radial or chamfer dimension to prevent the lead wire from changing the direction side.	Press <Ctrl> while dragging the dimensions
Quadrant for measuring the size of freezing angle	Right click and select the lock angle and place the dimension
To quickly change the additional text box of activities in the screen display text input box when creating or editing dimensions	Right click and select Edit additional text
Quick change of decimal digit displayed in dimension value	Any numeric key in the range of <Alt> + 0 to 6
Quickly add the broken line to the branch line of the coordinate size.	(1) For new dimensions, hold the <Shift> key while placing the dimension. (2) For the existing size, hold down the <Shift> key and drag the dimension

Continued

Objective	Operation
Aligning two coordinate sizes and relocating them simultaneously	Press the <Ctrl> key and drag one size to the second until it highlights, then click to place the two sizes
Enter edit mode	(1) For new dimensions, size is placed when clicking the middle key (2) For existing dimensions, double click size or right click size and select Edit
Exit edit mode	Perform one of the following operations: (1) In the size format box, click Edit again (2) Click the middle button (3) In the graphics window, right-click and select Edit

7.3.4 PMI hole callouts

Because the associated object for a PMI hole callout is the hole feature, rather than the geometry of the hole itself, the following points should be considered when creating and reusing a PMI radial or linear hole callout dimension.

(1) You have to select the hole feature itself when creating a PMI hole callout. Therefore, the callout must reside in the part that contains the feature. You can use PMI assembly filters if you need to display the PMI hole callout from a component in the context of an assembly.

(2) While most PMI annotations can be freely stacked, the PMI hole callout dimension cannot be added to a stack because doing so would break the necessary association with the hole feature. Other PMI annotations can be stacked onto a hole callout. In this case, the stacked PMI inherits an association with the hole feature faces, and not the hole feature itself.

(3) The *Find PMI Associated to Geometry* command only returns a PMI hole callout if the associated hole feature is selected. If you select a face from the feature, instead of the feature itself, the PMI hole callout will not be returned unless the face itself has been identified as an associated object.

(4) PMI hole callouts cannot be WAVE linked. This is because features are not WAVE linked, only bodies or topological entities can be WAVE linked.

(5) PMI hole callouts can not be mirrored. This is because only a body can be mirrored; a feature cannot be mirrored.

(6) If add itional model edits have affected the parameters of a hole callout, the hole callout will reflect the actual state of the hole feature, and not the original modelled state of the feature.

(7) Additional hole callout limitations may apply if a pattern of holes is created with a single hole feature. For example, multiple holes created from points in a single sketch.

7.4 Annotations

7.4.1 Note

Use *Note* to create and edit notes and labels. Select *PMI* tab > *Note* (Figure 7-7). A note consists of text, while a label consists of text with one or more leader lines. Text can be imported by reference to expressions, part attributes and object attributes, and can include symbols formed from control character sequences or user-defined symbols. While editing or creating notes, labels, or GD&T, NX provides a preview directly in the graphics window as you enter each character.

7.4.2 Feature Control Frame

Use *Feature Control Frame* to specify the geometric tolerances applied to the model features. Select *PMI* tab > *Feature Control Frame* (Figure 7-8). You can create and edit the control boxes with or without leader similar to drafting GD&T.

Figure 7-7 *Note* dialog box   Figure 7-8 *Feature Control Frame* dialog box

7.4.3 Datum Feature Symbol

A datum feature symbol indicates a datum feature on a part. Datum feature symbols are referenced from feature control frames to specify how a part is positioned during manufacturing and inspection. Select *PMI* tab > *Datum Feature Symbol*. The Figure 7-9 shows three types of leaders commonly used for datum feature symbols.

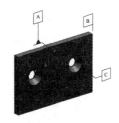

Figure 7-9 Leaders used for datum feature symbols

7.4.4 Datum Target

The *Datum Target* command can create a datum target symbol on a part to indicate a point, line, or area specific to a datum on the component. Select *PMI* tab > *Datum Target* (Figure 7-10). The datum target symbol is a circle divided into two parts. The lower part contains the reference letter and the reference target number. For the area type datum target, the identifier can be placed in the upper half of the symbol to show the shape and size of the target area.

7.4.5 Surface Finish

The *Surface Finish* command can create standards-compliant surface finish symbols on model faces to designate surface textures of solid materials. Select *PMI* tab >*Surface Finish* (Figure 7-11). Surface finish symbols can be either associative or non-associative to model geometry. They can be associative to linear model geometry such as edges, silhouettes, and section edges. They can also be made associative to dimensions and centerlines. You can create both associated and non-associative surface finish symbols with or without leader lines.

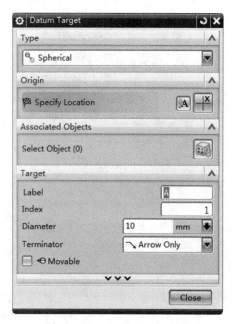

Figure 7-10 *Datum Target* dialog box   Figure 7-11 *Surface Finish* dialog box

7.4.6 Weld Symbol

The *Weld Symbol* command can create all kinds of welding symbols in metric and British parts and drawings. Select *PMI* tab >*Weld Symbol* (Figure 7-12). Welding symbol belong to correlative symbol, which can be replaced when the associated geometry changes or is marked out of date. You can edit the welding symbol properties, such as text size, font, scale and arrow size.

7.4.7 Balloon

The *Balloon* command can reference some other piece of information documented in another portion of the model. For example, parts lists, hole charts, and other tabulated formats often use balloon notes to directly reference geometry in the part whose content is in expanded format in a table. Select *PMI* tab >*Balloon* (Figure 7-13).

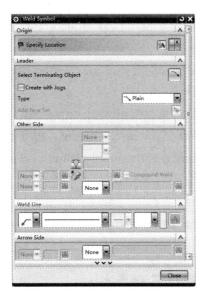 

Figure 7-12　*Weld Symbol* dialog box　　　　Figure7-13　*Balloon* dialog box

7.5　Section View

Use the *PMI Section View* command to create a section of all parts or selected parts in your assembly so that you can apply PMI to the sectioned geometry. Select *PMI* tab > *Section View* (Figure 7-14).

In the *Section View* dialog box, *Type* can specify the type of *PMI Section View*:
- One Plane: A single clipping plane.
- Two Parallel Planes: A pair of clipping planes, constrained to be parallel.
- Box: Three pairs of planes, locked ortho-normal to each other to form a cube.

Objects to Section can select the objects to be clipped.

Section Name can set the name for the PMI Section View as it appears in the *Part Navigator* list of model views.

Figure 7-14　*Section View* dialog box

Section Plane can set the location and orientation of the section plane.

Offset can specify the distance of the section plane from the coordinate system origin.

Display Settings>Show Manipulator can display the manipulator, which you can use to rotate the plane, or to change the offset distance.

Display Settings>Move Manipulator in View can move the manipulator to the center of the view.

Display Settings>Orient View to Plane can orient the model in the graphics window so that the section plane is parallel to the graphics window.

Cap Settings>Show Cap can turns on visibility of the cap area of the plane, which is where the section plane intersects the geometry of your model.

Cap Settings>Color Option can specify the method for coloring cap areas.

Cutting Plane Symbol > Display Cutting Plane Symbol can create a PMI cutting plane symbol for the section view. You can control cutting plane symbol visibility in Model Views as with other PMI objects.

Cutting Plane Symbol > Text can specify the text for the cutting plane symbol label.

Cutting Plane Symbol > Font can specify the line pattern for the border of the cutting plane symbol.

7.6 Exercise-PMI Example

7.6.1 Add PMI to a 3D model

In this exercise, product and manufacturing information will be added to the 3D model.

(1) Open the exercise *ch7\example\7-1.prt*.

(2) In the *Part Navigator* dialog box, set *Front View* as the work view, and rotate parts to view each aspect (as shown in Figure 7-15).

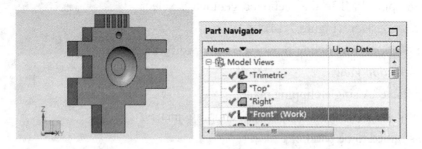

Figure 7-15 Part model

(3) In the *PMI* tab, select *Datum Feature Symbol*. Create a guide line to connect to the reference feature symbol. In the *Leader* group, click *Select Terminating Object* (as shown in Figure 7-16).

(4) Select the surface as shown in Figure 7-17, place the datum feature symbol-*A* in the

graphics window, and then click *Close*.

Figure 7-16   *Datum Feature Symbol* dialog box   Figure 7-17   Datum feature symbol-A

(5) In the *Part Navigator* dialog box, set *Left View* as the work view, create the datum feature symbol B and C (as shown in Figure 7-18).

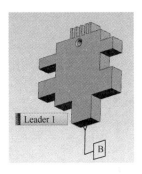 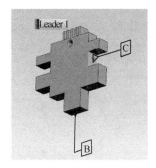

Figure 7-18   Datum feature symbol B and C

(6) In the *PMI* tab, select *Feature Control Frame*.

(7) In the *Feature Control Frame* dialog box, select *Position* as *Characteristic*, select diameter symbol ϕ for tolerance region, input 0.2 mm as tolerance, select Maximum material condition $Ⓜ$ (as shown in Figure 7-19).

(8) In the *Feature Control Frame* dialog box, set *Primary Datum Reference as A*, set *Secondary Datum Reference as B*, set *Tertiary Datum Reference as C* (as shown in Figure 7-20).

 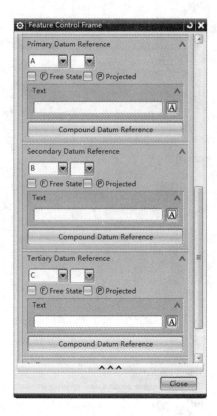

Figure 7-19　*Feature Control Frame* dialog box　　　Figure 7-20　*Feature Control Frame* dialog box

(9) In the *Associated Objects* group, click *Select Object*, and choose the face of the hole as shown in Figure 7-21.

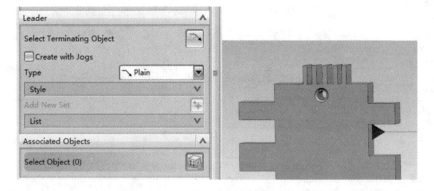

Figure 7-21　*Select Obiect*

(10) In the *Leader* group, click *Select Terminating Object*, and choose the edge of the hole as shown in Figure 7-22.

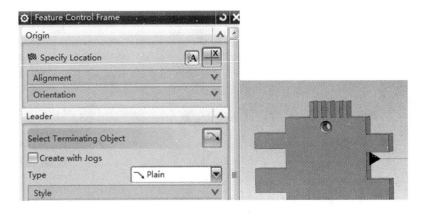

Figure 7-22 *Select Terminating Object*

(11) Place the feature control frame in the graphics window (as shown in Figure 7-23), and then click *Close*.

(12) In the *PMI* tab, select *Linear*.

(13) In the *References* dialog box, choose the face and the hole as shown in Figure 7-24 as the objects.

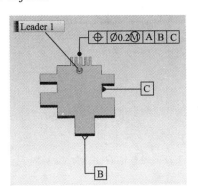 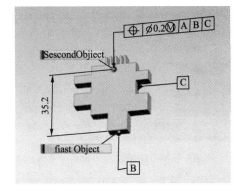

Figure 7-23 Place the feature control frame   Figure 7-24 Choose the face and the hole

(14) Hold the cursor in the graphics window until the new window appears (as shown in Figure 7-25), choose *Basic* in the *Size Display Type* option, click to place the dimension.

(15) In the *Part Navigator* dialog box, check the *Model View* to see the listing of PMI, and choose MPI and use correlation menu to display MPI (as shown in Figure 7-26).

(16) In the *Display in Views* dialog box, select *Trimetric* and click *OK* (as shown in Figure 7-27).

(17) In the *Part Navigator* dialog box, set *Trimetric View* as the work view (as shown in Figure 7-28).

(18) Exports the contents of the displayed part to a JT file, choose to include PMI.

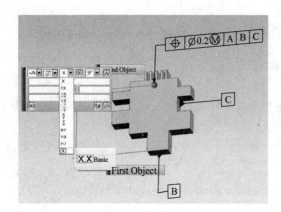

Figure 7-25   Size display type 　　　　　　Figure 7-26   Listing of PMI

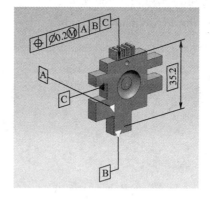

Figure 7-27   Display in Views　　　　　　Figure 7-28   PMI in Trimetric View

7.6.2   Create a PMI section view

This example shows how to create a PMI section view with one plane.

(1) Open the exercise ch7\example\7-2.prt, as shown in Figure 7-29.

(2) Select *PMI* tab >*Section View*.

(3) From the *Type* list, select *One Plane*.

(4) In the Section Plane group, click *Set Plane to Z* (as shown in Figure 7-30).

(5) Drag the handles to change the location or the orientation of the section (as shown in Figure 7-31).

(6) Click *OK* to create the section view (as shown in Figure 7-32).

Figure 7-29 Part model

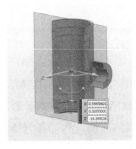

Figure 7-30 Plane to Z

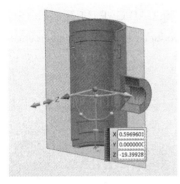

Figure 7-31 Drag the handles

Figure 7-32 Section view

7.7 Questions and Exercises

1. Fill in the blanks

(1) The_____option lets you explicitly capture your design intent by directly specifying the portions of the model to which the PMI pertains.

(2) NX supports the creation of standard dimension types, including baseline, _____ and_____.

(3) You can convert a feature dimension to a PMI feature dimension using the_____ option in the_____dialog box.

(4) _____are referenced from feature control frames to specify how a part is positioned during manufacturing and inspection.

(5) The_____command can create standards-compliant surface finish symbols on model faces to designate surface textures of solid materials.

(6) Use the 'PMI Section View' command to create a section of all parts or selected parts in your assembly so that you can apply_____to the sectioned geometry.

2. Short answer questions

(1) Briefly describe the benefits of PMI.

(2) Briefly describe the general process of adding PMI to a 3D model.

(3) Briefly describe the method of creating a PMI section view.

References

[1] 宁汝新，赵汝嘉. CAD/CAM 技术[M]. 北京：机械工业出版社，1999.

[2] 童秉枢等. 机械 CAD 技术基础[M]. 北京：清华大学出版社，1996.

[3] 王贤坤. 机械 CAD/CAM 技术、应用与开发[M]. 北京：机械工业出版社，2000.

[4] 洪如瑾. NX7 CAD 快速入门指导[M]. 北京：清华大学出版社，2011.

[5] 张瑞亮. 三维机械设计基础基础[M]. 北京：国防工业出版社，2013.

[6] 丁华. 三维机械设计基础基础[M]. 北京：科学出版社，2016.

[7] CHANG KUANGHUA. Product Design Modeling using CAD/CAE [M]. Elsevier Science, 2014.

[8] Siemens NX 12.0 Help Document [CP]. Siemens Product Lifecycle Management Software Inc., 2018.